Progress in Mathematics

Volume 71

Séminaire de Théorie des Nombres, Paris 1985–86

Edited by
Catherine Goldstein

1987

Birkhäuser
Boston · Basel

Catherine Goldstein
Université de Paris-Sud
Centre d'Orsay
F-91405 Orsay Cedex
France

"The Library of Congress has cataloged this
serial publication as follows:"
Séminaire Delange-Pisot-Poitou.
Séminaire de théorie des nombres / Séminaire Delange-Pisot-Poitou. — 1979–80- — Boston:Birkhäuser, 1981–
v. ; 24 cm. — (Progress in mathematics)
Annual.
English and French.
Continues: Séminaire Delange-Pisot-Poitou. Séminaire Delange-Pisot-Poitou : [exposés]
1. Numbers, Theory of—Periodicals. I. Title. II. Series: Progress in mathematics (Boston, Mass.)
QA24.S37a 512'.7'05—dc19 85-648844
Library of Congress [8510] AACR 2 MARC-S

CIP-Kurztitelaufnahme der Deutschen Bibliothek
Séminaire de Théorie des Nombres:
Séminaire de Théorie des Nombres. — Boston ; Basel :
Birkhäuser
Teilw. auf d. Haupttitels. auch: Séminaire Delange-Pisot-Poitou
1985/86. Paris 1985–86. — 1987.
(Progress in mathematics ; Vol. 71)
ISBN 3-7643-3369-3 (Basel)
ISBN 0-8176-3369-3 (Boston)
NE: GT

ISBN 0-8176-3369-3
ISBN 3-7643-3369-3

Text prepared by the editor in camera-ready form.
Printed and bound by Edwards Brothers Incorporated, Ann Arbor, Michigan.
Printed in the U.S.A.

9 8 7 6 5 4 3 2 1

Ce nouveau volume, concocté par la R.C.P. 303 du C.N.R.S. qui organise le Séminaire de Théorie des Nombres de Paris, rassemble des textes écrits par des conférenciers invités à ce séminaire pendant l'année 1985/86. Sans avoir la prétention de couvrir tous les aspects de la Théorie des Nombres, il s'efforce seulement de porter témoignage de la variété de la discipline et de ses renouvellements.

Je tiens bien sûr à remercier de leur aide tous les participants réguliers ou non du séminaire et tous ceux qui ont accepté de réviser les textes qui suivent. Mais c'est surtout grâce à la collaboration de Monique Le Bronnec qui a assuré la frappe définitive de tous les manuscrits et le travail du secrétariat que ce volume a pu être réalisé : je suis heureuse de trouver ici l'occasion de lui exprimer ma reconnaissance.

C. GOLDSTEIN

CONTENTS

Ce livre reproduit la plupart des conférences faites au Séminaire de Théorie des Nombres de Paris, 1985-86.

COMPUTING STABLE REDUCTIONS

Robert COLEMAN

Sections 1-5 below form a description of joint work with W. McCallum [C-M]. Section 6 contains a sketch of an algorithm for computing the stable reduction of a cyclic p-covering of $\mathbb{P}^1$. Section 7 provides an example to which this algorithm is applied.

1. Introduction. Let m be a positive integer and μ_m the group of mth roots of unity in $\overline{\mathbb{Q}}$, an algebraic closure of $\mathbb{Q}$. Let $K=\mathbb{Q}(\mu_m)$, $\mathcal{O}$ the ring of integers in K, and let $\wp$ be a prime of K. Let $\mathbb{F}_\wp$ denote the residue field at $\wp$ and suppose $\text{char}(\mathbb{F}_\wp)=p$. For $\psi : \mathbb{F}_p^\times \longrightarrow \mu_p$ a character and $a\in\mathbb{Z}$, such that $p \nmid \frac{m}{(m,a)}$ we set

$$G_a^{(m)}(\psi,\wp)= - \sum_{x\in\mathbb{F}_\wp^*} (\tfrac{x}{\wp})_m^a\, \psi(\text{Trace}_{\mathbb{F}_\wp/\mathbb{F}_p}(x))$$

where $(\frac{x}{\wp})_m^a : \mathbb{F}_\wp^* \longrightarrow (\mu_m)^a$ is the power residue symbol characterized by

$$(\tfrac{x}{\wp})_m^a \equiv x^{\frac{(q-1)a}{m}} \mod \wp.$$

For integers such that $a+b+c=0$, and $\wp$ such that $(\wp,m)=1$, we set

$$J_{a,b,c}(\wp)= \frac{1}{1-N\wp} \sum_{\substack{x+y+z=0\\ \in \mathbb{F}_\wp}} (\tfrac{x}{\wp})_m^a\, (\tfrac{y}{\wp})_m^b\, (\tfrac{z}{\wp})_m^c$$

$$= -(\tfrac{-1}{\wp})_m^c \sum_{x\in\mathbb{F}_\wp} (\tfrac{x}{\wp})_m^a\, (\tfrac{1-x}{\wp})_m^b.$$

We also have

$$J_{a,b,c}^{(m)}(\mathfrak{p}) = \frac{G_a^{(m)}(\mathfrak{P},\mathfrak{p})G_b^{(m)}(\mathfrak{P},\mathfrak{p})G_c^{(m)}(\mathfrak{P},\mathfrak{p})}{\mathbb{N}\,\mathfrak{p}}.$$

Fix m and drop it from the notation. Now let $\mathbb{I}_K$ denote the finite ideles of K. Then as Weil [W-1] has observed there exists a unique continuous (with respect to the natural topology on $\mathbb{I}_K$ and the discrete topology on K^*) character

$$\alpha : \mathbb{I}_K \longrightarrow K^*,$$

satisfying

(i) If $s \in \mathbb{I}_K$ $(\mathrm{supp}(s),m) = 1$,

$$\alpha(s) = \prod_{\mathfrak{p}} J_{a,b,c}(\mathfrak{p})^{\mathrm{ord}_{\mathfrak{p}} s}.$$

(ii) If $s \in K^* \subseteq \mathbb{I}_K$,

$$\alpha(s) = s^{\Phi_{a,b,c}}$$

where

$$\Phi_{a,b,c} = \sum_{t=1}^{m}{}' \left[\langle \frac{at}{m} \rangle + \langle \frac{bt}{m} \rangle + \langle \frac{ct}{m} \rangle \right] \sigma_{\epsilon t}^{-1} \in \mathbf{Z}[\mathrm{Gal}(K/\mathbf{Q})]$$

where $\epsilon=(-1)^{r}$, $r' = \langle \frac{a}{m} \rangle + \langle \frac{b}{m} \rangle + \langle \frac{c}{m} \rangle$ and where $\zeta^{\alpha_t}=\zeta^t$ for $\zeta\in\mu_m$ (for $r\in\mathbf{R}$, $\langle r\rangle\in[0,1)$ such that $r-\langle n\rangle\in\mathbf{Z}$), and " ' " indicates the summation is taken over indices prime to m).

As a consequence we obtain, for each prime $\mathfrak{p}$ of K (dividing m or not) a continuous character $\alpha_{\mathfrak{p}}$ such that

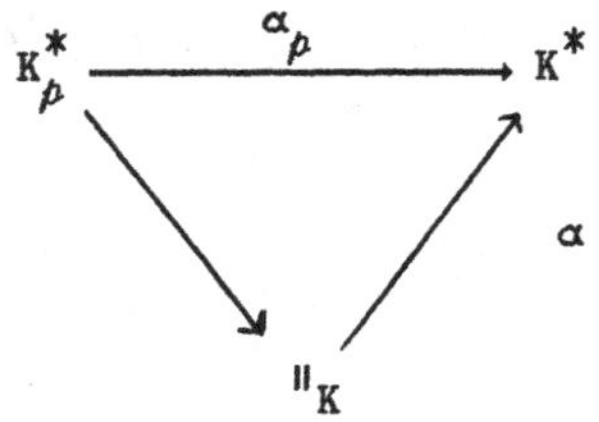

commutes. We know $\alpha_{\mathfrak{p}}$ completely when $\mathfrak{p}\nmid m$;

$$\alpha_{\wp}(x) = J_{a,b,c}(\wp)^{\mathrm{ord}_{\wp} x}.$$

For general $\wp$ we know by continuity there is a largest ideal $f_{\wp}$ in $\mathcal{O}_{\wp} \subseteq K_{\wp}$ such that

$$\alpha_{\wp}(1+f_{\wp}) = 1.$$

The ideal $f_{\wp}$ is called the conductor of α at $\wp$. In 1952, Weil [W-1] showed

$$f_{\wp} \geq m^2 \mathcal{O}_{\wp}$$

but observed that it could be much bigger. He raised the question of explicitly determining $f_{\wp}$. This question has since been investigated by several authors; Hasse in 1959 [H], Jensen in 1960 [J], Schmidt in 1980 [S] and Rohrlich (in press) [R]. Hasse computed $\alpha_{\wp}$ and $f_{\wp}$ precisely, when m is prime. The other authors gave estimates for, and in some cases determined, $f_{\wp}$. We have obtained precise formulas for both $\alpha_{\wp}$ and $f_{\wp}$ so long as $\wp \nmid 2$. We shall describe our formula for $\alpha_{\wp}$ $(\wp \nmid 6)$ below.

2. Statement of Results.

Let m,p and a,b,c be as above. Suppose $p|m$ and $(m,a,b,c)=1$. Set e equal to 1, if $(p,abc)=1$, and equal to the unique integer in $\{a,b,c\}$ divisible by p, otherwise. For $x \in K_{\wp}^*$, let σ_x denote the image of x in $G_{K_{\wp}}^{ab}$ under the Artin map. Let $\pi \in \mathbb{Q}_p(\mu_p)$ such that $\pi^{p-1}=-p$.

Theorem 1. Suppose $p \geq 5$ or $p=3$ and $e=1$. Let $n= \mathrm{ord}_p m$. Then for $x \in \mathcal{O}_{\wp}^*$

$$\alpha_{\wp}(x) = \begin{cases} (\pi p^n)^{\frac{e(\sigma_x - 1)}{m}} & p \nmid \frac{m}{(m,e)} \\[2ex] (\pi^m, a^a b^b c^c)_2^{\frac{\sigma_x - 1}{2m}} & \text{otherwise.} \end{cases}$$

Now let ζ be a primitive p^nth root of unity in $K_{\wp}$ and for $t \in \mathbb{F}_p$ let

$$\varphi(t) = (\varsigma^{-p^{n-1}})^t.$$

Finally, let $\beta = \xi^{-1/2} - \xi^{1/2}$, then

$$\alpha_{\wp}(\beta) = \begin{cases} \left((-1)^c\left(\frac{a}{b}\right)^a\right)^{\frac{\sigma_\beta - 1}{m}} G_a^{(m)}(\varphi,\wp) & p \nmid \frac{m}{(m,e)} \text{ and } e=a \\ (-2mabc)^{\frac{\sigma_\beta - 1}{2}} (a^a b^b c^c)^{\frac{\sigma_\beta - 1}{m}} G_1^{(2)}(\varphi,p) & \text{otherwise.} \end{cases}$$

Note that when m is odd, the terms involving $\frac{\sigma_\beta - 1}{m}$ are trivial.

<u>Corollary 1.1</u>. If $m=p$, $e=1$,

$$(\varsigma^{-1/2} - \varsigma^{1/2})^{\Phi_{a,b,c}} = (2abc)^{\frac{\sigma_\beta - 1}{2}} G_1^{(2)}(\varphi,p)$$

This formula was originally proven by Gauss in order to determine the sign of the quadratic Gauss sum.

<u>Remarks</u>. A formula for $\alpha_{\wp}$ when $p=3$ and $e>1$ may be obtained from the above theorem if one uses the Hasse-Davenport relation for Gauss sums. The determination of $f_{\wp}$ may be made from Theorem 1 using the explicit reciprocity law [C].

3. <u>Fermat Curves</u>.

Let F_m denote the smooth plane curve over $\mathbf{Q}(\mu_m)$ with homogeneous equation

$$X^m + Y^m + Z^m = 0.$$

Clearly $G_m = \mu_m^3/\Delta$ acts on F_m, where Δ is the image of μ_m in μ_m^3 under the diagonal embedding. The Jacobian J_m of F_m has CM by the image of the group ring $\mathbf{Z}[G_m]$ in $\mathrm{End}_K J_m$. This image is not an order in a number field so one must make quotients to apply the theory of Shimura-Tanayama. Let a,b,c be as above. Let

$$H_{a,b,c} = \frac{\{(\varsigma_1,\varsigma_2,\varsigma_3)\in\mu_m^3 : \varsigma_1^a\varsigma_2^b\varsigma_3^c = 1\}}{\Delta} \subseteq G_m$$

Let $F_{a,b,c} = F_m/H_{a,b,c}$. Then $F_{a,b,c}$ has the affine equation :

$$w^m = u^a(1-u)^b(-1)^c$$

and $G_m/H_{a,b,c}$ acts on $F_{a,b,c}$. The Jacobian A of $F_{a,b,c}$ has CM by the group ring $\mathbb{Z}[G_m/H_{a,b,c}]$.

Let f_m denote the mth cyclotomic polynomial. Let $J^{new}_{a,b,c}$ denote the quotient of A by the Abelian subvariety generated by $\{f_m(\sigma)A: (\sigma) = G_m/H_{a,b,c}\}$. We identify $G_m/H_{a,b,c}$ with μ_m via the map

$$\sigma \longrightarrow \sigma w/w.$$

Now $J^{new}_{a,b,c}$ has CM by $\mathcal{O}$. With notation as above, we deduce from the theory of Shimura-Tanyama the following :

<u>Theorem 2</u>. Suppose $t \in J^{new}_{a,b,c}(\overline{\mathbb{Q}})_{tor}$, $s \in II_K$ and $(\mathrm{supp}(s), \mathrm{order}(t))=1$, then

$$[s,K] = \alpha(s)t$$

where $[s,K] \in \mathrm{Gal}(K^{ab}/K)$ is the Artin symbol.

Hence to compute $\alpha_{\wp}$ we may use

<u>Corollary 2.1</u>. Let $t \in J^{new}_{a,b,c}(K_p)$. Then if $(\wp, \mathrm{order}(t))=1$, $\alpha_{\wp}(x)t = [x,K_{\wp}]t$.

Thus the computation of $\alpha_{\wp}$ may be carried out by studying the geometry of $F_{a,b,c} \times K_{\wp}$.

4. <u>Stable reduction</u>.

Let F denote a finite extension of $\mathbb{Q}_p$. Let $\mathcal{O}_F$ and $\tilde{F}$ denote the ring of integers and residue field of F. Suppose X is a smooth complete curve over F. We say that a scheme Y over $\mathcal{O}_F$ is a semi-stable model for X if

(i) Y is proper,

(ii) $Y \times_{\mathcal{O}_F} F = X$,

(iii) $\tilde{Y} \overset{defn}{=} Y \times_{\mathcal{O}_F} \tilde{F}$ has at worst normal crossings as singularities.

Suppose X has a semi-stable model Y over $\mathcal{O}$ such that $Y \times \mathcal{O}_{F'}$ is the unique minimal semi-stable model for $X \times F'$ over $\mathcal{O}_{F'}$ for each finite extension F' of F. Then Y is called a stable model for X over $\mathcal{O}_F$.

<u>Theorem 3</u> [D-M]. There exists a finite extension F' of F such that $X \times F'$ has a semi-stable model over $\mathcal{O}_{F'}$. Moreover, if the genus of X is at least 2 or X has genus one and $\mathcal{O}_F$-integral j-invariant, then $X \times F'$ has a stable model which is unique up to unique isomorphism.

Suppose Y is the stable model for X over $\mathcal{O}_{\overline{F}}$. Then if $\sigma \in G = G(\overline{F}/F)$, Y^{σ} is also a stable model for $X_{\overline{F}}$. Hence there exists a unique isomorphism

$$f_{\sigma} : Y^{\sigma} \xrightarrow{\sim} Y.$$

Suppose $\tilde{\sigma} = \phi^n$, $n \in \mathbb{N}$, where ϕ is the absolute Frobenius automorphism of $\mathbb{F}$ the residue field of $\overline{F}$. Then

$$\overline{Y^{\sigma}} = \widetilde{Y^{\sigma}} = \tilde{Y}^{\phi^n}.$$

Let $\mathrm{Frob} : Y \longrightarrow Y^{\phi}$ denote the absolute Frobenius morphism. Let

$$w(\sigma) = \overline{f^{\sigma}} \circ (\mathrm{Frob})^n : \tilde{Y} \longrightarrow \tilde{Y}.$$

This gives us a homomorphism of semigroups from

$$W^+(F) = \{\sigma \in G : \tilde{\sigma} = \phi^n : n \in \mathbb{N}\}$$

to $\mathrm{End}(\tilde{Y})$. We have

<u>Theorem 4</u>. Suppose $\mathrm{Jac}(X_{\overline{F}})$ has good reduction A. Then

(i) $A \cong \mathrm{Jac}(\tilde{Y})$ canonically;

(ii) $\mathrm{Jac}(X)'_{\mathrm{Tor}} \cong A'_{\mathrm{Tor}}$;

(iii) For $\sigma \in W^+(F)$, $t \in \mathrm{Jac}(X)'_{\mathrm{Tor}}$, $\sigma t = w(\sigma)\tilde{t}$

(here "'" reads prime to p).

Now we may apply this to Fermat curves since their Jacobians have CM and hence potential good reduction.

5. <u>Stable Reduction of Fermat Curves</u>.

For simplicity we will only discuss the special case $m=p^n$, $m \nmid e$. Then we have

<u>Theorem 5</u>. Let notation be as before. Let $r = \mathrm{ord}_p e$. Then $F_{a,b,c}$ attains stable reduction over the field

$$K(\sqrt{\pi}, \sqrt[p^n]{a^a b^b c^c})$$

and its stable reduction is :

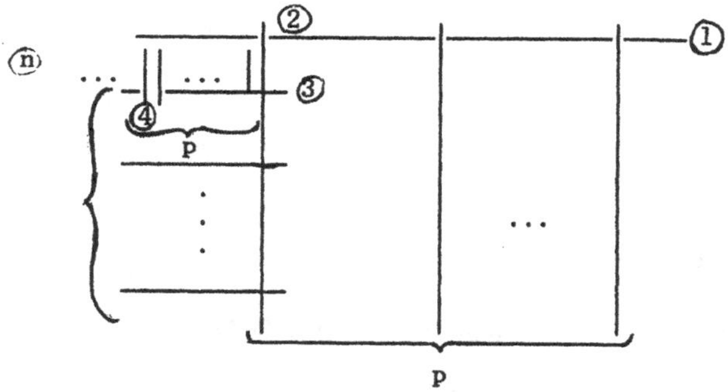

where the components on the first r levels have genus zero and those on the remaining levels are isomorphic to the curve with affine equation :

$$y^p - y = x^2.$$

Finally we need to describe $w(\sigma)$ for $\sigma \in \mathrm{Gal}(\overline{\mathbb{Q}}_p/\mathbb{Q}_p)$. For simplicity we suppose σ lies in inertia group of $\overline{\mathbb{Q}}_p/F$ where F is the unique unramified quadratic extension of $\mathbb{Q}_p$. Then let

$$\zeta_\sigma = \frac{\sigma(\sqrt[p^n]{a^a b^b c^c})}{\sqrt[p^n]{a^a b^b c^c}} \in \mu_{p^n},$$

$$\epsilon_\sigma = \frac{\sigma(\sqrt{\pi})}{\sqrt{\pi}} \in \mu_{2(p-1)}.$$

For more simplicity we suppose $\zeta_\sigma \in \mu_p$. Then σ stabilizes the components of the stable reduction and its action on a final component X after a suitable identification of X with

$$A_2 : y^p - y = x^2.$$

$w(\sigma)$ is determined by

$$w(\sigma) : (x,y) \longrightarrow (\epsilon_\sigma x, \epsilon_\sigma^2 y)$$

if $\zeta_\sigma = 1$, and

$$w(\sigma) : (x,y) \longrightarrow (x, y+1)$$

if $\epsilon_\sigma = 1$ and ζ_σ is a primitive p^{th} root of unity dependent on the identification of X with A_2.

Now μ_{p^n} acts on $F_{a,b,c}$ hence on its stable model, and ultimately on its stable reduction. The key to the computation of α_p is that if $\epsilon_\sigma = 1$, ζ_σ^1 acts on the stable reduction in the same way $w(\sigma)$ does.

6. Towards an Algorithm for Computing Stable Reductions.

Let X be a smooth complete curve over a finite extension F of $\mathbb{Q}_p$. It is known [D-M] that X achieves semi-stable reduction over the field F_ℓ of ℓ-torsion, $(\ell \neq p)$ of its Jacobian. Unfortunately, given equations for X, it is not so easy to find F_ℓ and hence to find a semi-stable model for X. Moreover, this is not the way the results described in the previous sections were obtained. They were actually obtained by some guesswork based on Hasse's results [H] and McCallum's thesis [M]. It would be desirable to have some efficient algorithm for computing the stable reduction of a curve. Below we describe an approach which succeeds in at least some cases.

First we define the notion of a stable model for a marked curve (X,S) with $S \subseteq X(\overline{F})$. It is the unique minimal semi-stable model Y for X over $\mathcal{O}_{\overline{F}}$ such that the map $\mathrm{red} : S \longrightarrow \tilde{Y}(\overline{\mathbb{F}})$ is an injection into the non-singular points of $\tilde{Y}$ ($\mathbb{F}$ is the residue field of F). It exists if $\#S \geq 3$ and $g(X)=0$, $\#S=1$ and $g(X)=1$ or $g(X) \geq 2$.

Now suppose $f : X \longrightarrow \mathbb{P}^1$ is a non-constant morphism. Let $S = f^{-1}\{\infty\}$. Suppose Y is the stable model for (X,S) over $\overline{F}$. Let

$\bar{\mathbb{F}}$ denote the residue field of $\mathbb{F}$. Let $\{Y_i\}$ be the set of irreducible components of $\tilde{Y}$. Let

$$Y_i' = Y_i - (S \cup \bigcup_{j \neq i} Y_j)$$

which is a Zariski open in $\tilde{Y}_i$. Let $\mathbb{Y}_i = \text{red}^{-1} Y_i'$, considered as an affinoid subdomain of X. Then one can show $f(\mathbb{Y}_i)$ is an affinoid subdomain of $A^1_{\mathbb{F}}$ with irreducible reduction. By the classification theorem for subdomains of $A^1_{\mathbb{F}}$ ([BGR]), $f(\tilde{Y}_i)$ must be of the form

$$B[a,r] - \bigcup_{b \in T} B(b,r)$$

where $a \in \mathbb{F}$, $T \subseteq \mathbb{F}$ is finite, $r \in |\mathbb{F}|$ and

$$B[c,s] = \{x \in F : |x-c| \leq s\}$$

$$B(c,s) = \{x \in \mathbb{F} : |x-c| < s\}.$$

Moreover, $Z = f^{-1}B[c,r]$ is an affinoide subdomain of $X-S$ whose reduction contains an irreducible component of $Y-\tilde{S}$.

For a one-dimensional reduced affine scheme V over an algebraically closed field K we define an integer $e(V)$ as follows. Let $\{V_i\}$ be the set of irreducible components of V. Let s denote the number of singularities on V counted with multiplicity. Let i denote the number of points at infinity of V in its complete non-singular model. Let n denote the number of connected components of V isomorphic to $A^1_{\bar{F}}$. Then

$$e(V) = (\sum_{V_i} (2g(V_i)-2)) + s + i + n.$$

One can show

<u>Lemma 6</u>. With notation as above, a closed disk $B \subseteq A^1_{\mathbb{F}}$ equals the union of $f(\tilde{Y}_i)$ with finitely many open disks for some integer i, iff $e(f^{-1}(B)^{\sim}) > 0$.

One approach toward finding Y is to find all closed disks $B \subseteq A^1_{\mathbb{F}}$ such that

$$e(f^{-1}(B)) \overset{\text{defn}}{=} e(f^{-1}(B)^{\sim}) > 0.$$

Unfortunately, at present there is no general algorithm for computing $f^{-1}(B)^{\sim}$. One useful general result, however, is

Lemma 7. Suppose $e(f^{-1}(B))>0$. Then if B if properly contained in a closed disk B', $B \subseteq f(\text{red}^{-1}s)$ where s is a singular point of $f^{-1}(B')^{\sim}$.

Examples. Suppose f is Galois of degree prime to p. Let $T \subseteq \mathbb{A}^1_{\bar{F}}$ denote the set of finite branch points of f. Then if B is a closed disk in $\mathbb{A}^1_{\bar{F}}$ such that $e(f^{-1}(B))>0$, B is of the form

$$B[a,(a-a')]$$

where $a, a' \in T$ and $a \neq a'$ (note that if $\#T \leq 1$ then X has genus zero and $\#f^{-1}(\infty)=1$, so $(X, f^{-1}(\infty))$ does not have a stable model). See [B] where the case $\deg f=2$ is discussed. Using this one can show that if f is Galois of degree prime to p and unbranched outside $\{0,1,\infty\}$ then X has potential good reduction. One can also use this to show that the power series expansion of an algebraic function converges up to the first branch point of its minimal polynomial as long as p does not divide the order of its Galois group. This implies the result of [D-R].

The situation becomes more complicated if f is Galois and $p \mid \deg f$. Below we will explain how to proceed when f is cyclic of degree p.

After adjoining the pth roots of unity we may assume X has the equation

$$y^p = r(x)$$

where $x=f$ and $r(x) \in F[X]$. Suppose $\deg r=n$ and let a, $\{a_i : 0 \leq i \leq [\frac{n}{p}]\}$, $\{b_i : 0 < i \leq n, (i,p)=1\}$ be variables. Let W denote the affine curve in $\text{Spec } F[a,\{a_i\},\{b_j\}]$ with equations codified by

$$g(x) = \Sigma\, a_i x^i$$

$$h(x) = \Sigma\, b_i x^i$$

$$r(x+a) = g(x)^p + h(x).$$

We may and will also think of $W(\bar{F})$ as contained in $\bar{F} \times \bar{F}[x] \times \bar{F}[x]$.

Suppose now $(a,g(x),g(x))\in W(\mathbb{F})$. Let $r, u_{a,r}\in|F|$ and $b,c\in F$ such that $|b|=r$ and

$$|c| = u_{a,r} = \max\{|\pi g(bx)|,\ |h(bx)|^{1/p}\}$$
$$= \max\{|\pi a_i|r^i,\ (|b_j|r^j)^{1/p}\}$$

where $\pi\in F$ such that $\pi^p=-p$. We set

$$g_{a,r} = \left(\frac{\pi g(b(x))}{c}\right)^{\sim} \in \mathbb{F}[x]$$

$$h_{a,r} = \left(\frac{h(bx)}{c^p}\right)^{\sim} \in \mathbb{F}[x].$$

We point out that these polynomials depend on the choices of b and c (as well as a and r), but not in a significant way.

Lemma 8. The ring $A^o(f^{-1}B[a,r])$ of integer valued rigid analytic functions on $f^{-1}B[a,r]$ is generated topologically by $z=\frac{y-g(bx)}{c}$ and x and its reduction is the curve

$$Y_{a,r}\ :\ z^p - (g_{a,r}(x))^{p-1}\, z = h_{a,r}(x).$$

It remains to determine when $e(Y_{a,r})>0$.

Lemma 9. Suppose $s(x), t(x)\in\mathbb{F}[x]$ such that either $s\neq 0$ or t is not a pth power and Z is the affine scheme determined by the equation

$$z^p - s(x)^{p-1}z = t(x).$$

Then Z is reduced and Z has all rational components iff one of the following holds :

(i) $s=0$

(ii) $t= u^p-s^{p-1}u+s^pL$, where $u(x)\in\mathbb{F}[x]$ and $L(x)\in\mathbb{F}(x)$ with $\deg L\leq 1$ or $L=0$.

Lemma 10. With notation as in the previous lemma, $e(Z)=0$ iff one of the following holds.

(i) $s=0$, and $f'(x)$ has only one zero.

(ii) $s\neq 0$ and has at most one zero, and $t= u^p-s^{p-1}u+s^pL$, where $u\in\mathbb{F}[x]$ and either s or L is constant.

<u>Corollary 10.1</u>. Suppose in addition to our previous assumptions that s is constant or $s(0)=0$ and f has no terms of degree divisible by p. Then $e(Z)=0$ iff one of the following holds :

(i) $s=0$ and $f'(x)$ has at most one zero.
(ii) $s\neq 0$ is constant and $\deg t=1$.
(iii) $s\neq 0$ and vanishes only at zero, and $t=0$.

We note that if B is an affinoid ball $\subseteq A^1_{\overline{F}}$, either there exists an $a\in B$ such that $r(a)=0$ and so $g_{a,r}(0)=0$ or $g_{a,r}$ is constant for all $a\in B$. Hence our task becomes, to determine all pairs $(a,r)\in\overline{F}\times|\overline{F}|$ such that either

(1.1) $g_{a,r}=0$ and $h_{a,r}$ has more than one zero; or
(1.2) $r(a)=0$ and $g_{a,r}\neq 0$ and has at least two zeros, or $h_{a,r}\neq 0$; or
(1.3) $g_{a,r}$ is constant, $g_{a,r}\neq 0$ and $\deg h_{a,r}>1$.

These conditions translate into the following inequalities among the coordinates of our point $(a,g(x),h(x))$. Either

(2.1) $|\pi^a|r^i < u_{a,r}$, $\forall\, i$ and there exist $i<j$, prime to p, such that $|b_i|r^i=|b_j|r^j=a^p_{a,r}$ and $i>1$ or $j\neq 1+p^k$; or
(2.2) $r(a)=0$ and $\exists\, i$ such that $|\pi a_i|r^i=u_{a,r}$ and either $\exists\, j\neq i$ such that $|\pi a_j|r^j=u_{a,r}$ or $\exists\, k$ such that $|b_k|r^k=u^p_{a,r}$; or
(2.3) $|\pi a_0|=u_{a,r}>|\pi a_i|r^i$ for all $i>0$. And there exists an $i>1$, $(i,p)=1$ such that $|b_i|r^i=u^p_{a,r}$.

One can eliminate r from these inequalities and only deal with inequalities among functions on W. For example, condition (iii) translates into $\left|\frac{\pi a_0}{b_i}\right|^{p/i}=r$ for some i and $|\pi a_0|\geq|\pi a_j|r^j$, $j>0$

$$|\pi a_0|\geq|b_j|r^j\ ,\quad j\geq 0.$$

Hence we must be able to solve the following sort of problem. Let

$$a : W \longrightarrow \mathbb{P}^1_{\overline{F}}$$

be a finite morphism of curves. Let ℓ be a function on W.

<u>Problem</u> : Describe explicitly $R=\{a(Q) : Q\in W(\overline{F}),\ |\ell(Q)|\leq 1\}$ as a subdomain of $\mathbb{P}^1_{\overline{F}}$.

By the classification theorem [BGR], if ℓ is non-constant, R is a finite union of connected subdomains of $\mathbb{P}^1_{\overline{\mathbb{F}}}$, each of which is the complement of a finite union of open disks in a closed disk. A solution to the problem amounts to giving centers and radii for these disks.

To solve this problem, let $g_a(X) = c_m(a)X^m + \ldots + c_0(a)$ be the minimal polynomial for ℓ over $\overline{\mathbb{F}}[a]$. Now $R-\{\infty\}$ is the set of all $a \in \overline{\mathbb{F}}$ such that the Newton polygon of $g_a(X)$ has a slope ≤ 0. This imposes certain inequalities among the polynomials $c_i(a)$ and hence we are reduced to the following general problem. Let $T(a) \in \overline{\mathbb{F}}(a)$.

<u>Problem</u> : Describe explicitly the set $\{a : |T(a)| \leq 1\}$.

The solution of this problem is implicit in the proof of the classification theorem. Its solution can and should be made more explicit; however, we have not had the time to do this. In the next section we will apply this algorithm to a particular example.

7. <u>The Stable Reduction of</u> $F^9_{3,1} : w^9 = u^3(1-u)$.

Note that this is an example of a curve not covered by Theorem 5. Set $x = w/u$. Then $F^9_{3,1}$ has the equation

$$w^3 = x - x^4 \overset{\text{def}}{=} r(x)$$

hence is a cubic covering of $\mathbb{P}^1_{\mathbb{Q}_3}$ so we may apply our algorithm. Let

$$\begin{aligned} g(x) &= a_0 + a_1 x \\ h(x) &= b_1 x + b_2 x^2 + b_4 x^4 \end{aligned}$$

with the relations

$$r(x+a) = g(x)^3 + h(x).$$

This entails the equations

$$\begin{aligned} a_0^3 &= a - a^4 \\ a_1^3 &= -4a \\ b_1 &= -(3a_0^2 a_1 + 4a^3 - 1) \\ b_2 &= -(3a_0 a_1^2 + 6a^2) \\ b_4 &= -1\ . \end{aligned} \tag{1}$$

Let $|b|=r$, $|c|= r_{a,r}= \max \{|\pi g(bx)|, |h(bx)|^{1/3}\}$. Let $g_{a,r}$ and $h_{a,r}$ be as in the previous section. We now conduct the search for the closed disks B such that $e(w^{-1}(B))>0$.

Let $Y_{a,r}= (w^{-1}B[a,r])^{\sim}$. We observe that $g_{0,r}=0$ and $b_1=1$, $b_2=0$, $b_4=-1$.

$$u_{0,r} = \max \{r,r^4\} = \begin{cases} r & \text{if } |r| \le 1 \\ r^4 & \text{if } |r| > 1 \end{cases}.$$

So $Y_{0,r}$: $\begin{cases} z^3= x, & |r| < 1 \\ z^3= x-x^4, & |r| = 1 \\ z^3= x^4, & |r| > 1 \end{cases}$.

So $e(Y_{0,r})=0$ and has a singularity at $x=0$ if $|r|>1$, at $x=1$ if $|r|=1$ and no singularity if $|r|<1$. Using Lemma 7 we conclude that if $e(Y_{a,r})>1$ then $|a-1|<1$ and $r<1$. We now apply the algorithm as explained in the previous section, taking into account the information acquired above.

<u>Case (i)</u>. $g_{a,r}=0$ and $h'_{a,r}$ has at least two zeros. This implies $u_{a,r} > \max \{|\pi a_0|, |\pi a_1|r\}$, $u^3_{a,r}= r^4= |h|r^2 \ge |b_1|r$. We know $|a-1|<1$ so $|a|=1$, $|a_0|^3= |a^3-1|<1$, $|a_1|=1$, $|b_2|=|3|$, $|b_1| \le \max\{3, |a_0|^3\}$. We deduce $r^2= |3|$ so $|r|=|\pi|$ and

(2) $$|\pi|^3 \ge |b_1| = |3a_0^2a_1 + 4a^3 - 1|.$$

This implies $|a_0|^3 = |3|$. This is clearly satisfied if

(3) $$3a_0^2a_1 + 4a^3 -1 = 0.$$

Moreover, if a, a_0, a_1 is a solution of (2) and (3), and a', a_0', a_1' satisfies (1) and (2), then $|a-a'| \le |\pi|$. Hence we obtain at most one closed disk B with $e(w^{-1}(B))>0$ from case (i). Finally, (3) implies a is a root of

$$f(z)= (4z^3-1)^3 + 108z(z-z^4)^2.$$

Conversely, if a is a root of $f(z)$ we can find a_0, a_1 so that (1) is satisfied as well as (2), and $Y_{a,|\pi|} : z^3 = x^2 + x^4$.

<u>Case (ii)</u>. $r(a)=0$, $g_{a,0}\neq 0$.

Then $h_{a,r}\neq 0$, $a^3=1$ so $a_0=0$, $|a_1|=1$, $|b_1|=|3|$; $|b_2|=|3|$ and $u_{a,r}=|\pi|b$. $u^3= \max\{|3|r, |3|r^2, r^4\}$ which is impossible.

<u>Case (iii)</u>. $g_{a,r}\neq 0$ is constant and $\deg h_{a,r}>1$. Thus

$$u_{a,r} = |\pi a_0| > |\pi a_i| r , \quad \text{and}$$
$$u^3_{a,r} = \max\{|b_2|r^2, r^4\} \geq |b_1| r .$$

Also we have $|a^3-1|<1$ so that $|a_0| = |a^3-1|^{1/3}<1$, $|a_1|=|a|=1$, $|b_2|=|3|$.

(a) Suppose $u^3_{a,r}=r^4$. Then $r^4 \geq r^2|b_2|$ becomes $|\pi a_0|^{3/2} \geq |3|$ which implies $|a_0|^3 \geq |\pi|$. $|a_0|^3= |1-a^3| \geq |\pi|$ so that $|b_1|>|1-a^3|= |a_0|^3$ but then $r^4 \geq |b_1|r$ becomes $|\pi a_0|^9 \geq |a_0|^{12}$ or $|\pi|^9 \geq |a_0|^3 \geq |\pi|$, a contradiction.

(b) $|u_{a,r}|= |b_2|r^2 > r^4$. We have $|b_2|r^2= |\pi a_0|^3$ which implies $r= |\pi a_0^3|^{1/2}$ so that $|b_1|r^2 \geq |b_1|r$ becomes

$$(4) \qquad |3||\pi a_0^3|^{1/2} \geq |3a_0^2 a_1 + 4a^3 - 1|.$$

This implies $|a_0|^3 = |3|$ using $a_0^3= a-a^4$ and $r= |3\pi|^{1/2}$. Once (4) holds all the other inequalities hold. Rewriting (4) we must have $|27\pi|^{1/2} \geq |3a_0^2a_1+4a^3-1|$ and $|a_0|^3=|3|$.

If a, a_0, a_1 are such that

$$(5) \qquad 3a_0^2a_1 + 4a^3 - 1 = 0$$

then $Y_{a,r} : z^3- z = x^2$ which has genus 1. As in case (i) this means $f(a)=0$. We conclude $e(W-B)>0$ iff $B= B[a,r]$ where a is a root of $f(z)$ and $r= |3\pi|^{1/2}$ or $r= |\pi|$.

Note that if a is one root of $f(z)$, so is ζa, $\zeta \in \mu_3$, and if $\zeta \neq 1$, $\zeta a \notin B[a, |3\pi|^{1/2}]$. Hence we find at least three closed disks from case (iii). It is not hard to check that these are all. From this information one can conclude that $F^9_{3,1}$ has stable reduction over the splitting field of $f(z)$ and its stable reduction is :

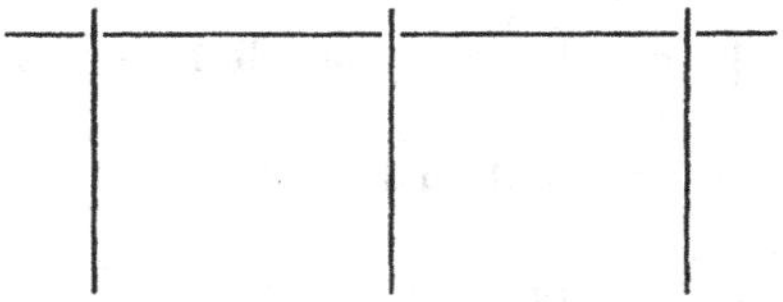

where the horizontal component has genus zero and the vertical components are isomorphic to

$$z^3 - z = x^2.$$

See [C-M] for additional remarks helpful in drawing these final conclusions.

Final Remarks.

1. We have been able to use recent results of Ihara [I] and Anderson [A-2] to give an independent proof of the formula for the wild part of $\alpha_{\mathfrak{p}}$ even when $\mathfrak{p}|2$. We expect that formulas for the local components of more general Jacobi Sum Hecke characters [A-1], [K-L], [W-2] can be found using Anderson's results [A-2].

2. It seems likely that the algorithm sketched in sections 6 and 7 can be generalized to cyclic coverings of arbitrary degree. It is an interesting problem to find a general algorithm. (See [B-L] and [VP] for rigid analytic proofs of stable reduction. Although these proofs are ineffective, it was the rigid analytic approach which led us to the results described in this paper).

3. A curve C of genus 2 has seven possible types of stable reduction. If the curve has the equation $y^2=f(x)$ for a polynomial f of degree 5 or 6 it should be possible to give a nice criterion (analogous to the integrality of the j-invariant) based on the coefficients of f for the type of stable reduction C has.

BIBLIOGRAPHY

[A-1] . Anderson.- Cyclotomic and a covering of the Taniyama group.

[A-2] G. Anderson.- The hyperadellic gamma function (to appear).

[B] F.A. Bogomolov.- Sur l'algébricité des représentations χ-adiques, C.R. Acad. Sci. 290 (1980).

[Bo] S. Bosch.- Formelle standard modellehyperelliptischer Kurven, Math. Ann. 251 (1980), 19-42.

[BGR] S. Bosch, U. Guntzer and R. Remmert.- Non-Archimedian Analysis (Springer-verlag, Berlin, 1984).

[B-L] S. Bosch and W. Lutkebohmert.- Stable reduction and uniformization of abelian varieties, I, Math. Ann.

[C] R. Coleman.- The dilogarithm and the norm residue symbol, Bull. Math. Soc. France 109 (1981), 373-402.

[C-M] R. Coleman and W. McCallum.- Stable reduction of Fermat curves and Jacobi sum Hecke characters, to appear.

[D-M] P. Deligne and D. Mumford.- The irreducibility of the space of curves of a given genus, Publ. Math. IHES n° 36 (1969), 75-109.

[D-R] B. Dwork and P. Robba.- On Natural R-adic of p-adic convergences. Trans. AMS, 256 (1979), 199-213.

[H] H. Hasse.- Zeta Funktion und L-funktionen zu einem arithmetishen funktionen - Körper vom fermatschen Typus, Abhand. der Deut. Akad. der Wissen zu berlin (1955).

[J] C. Jensen.- Über die Führer einer Klasse Heckeschen Grössencharakter, Math. Scand. 8 (1960), 151-165.

[K-L] D. Kubert and S. Lichtenbaum.- Jacobi sum Hecke characters, Compositio Math. 48 (1983), 55-87.

[K] D. Kubert.- Jacobi sums and Hecke characters, Amer. J. Math. 107, N° 2 (1985), 253-280.

[R] D. Rohrlich.- Jacobi sums and explicit reciprocity laws, to appear in Comp. Math.

[S] C.-G. Schmidt.- Über die Fuhrer von Gausschen Summen als Grössencharactere, J. Number Theory 12 (1980), 283-310.

[V] M. Vanderput.- Stable reductions of algebraic curves, to appear.

[W-1] A. Weil.- Jacobi sums as Grössencharakter, Trans. Am. Math. Soc. 73 (1952), 487-495.

[W-2] A. Weil.- Sommes de Jacobi et caractères de Hecke, Gottingen Nachr. (1974), n° 1.

Robert Coleman
Department of Mathematics
University of California
Berkeley, California 94720
U.S.A.

RESULTATS RECENTS LIES AU THEOREME D'IRREDUCTIBILITE DE HILBERT

Pierre DEBES

Le théorème d'irréductibilité de Hilbert est un résultat de la fin du siècle dernier [17]. Le problème est le suivant : étant donnés un corps k et $P_1,\dots,P_n$ n polynômes irréductibles dans $k(X_1,\dots,X_r)[Y_1,\dots,Y_s]$, montrer que l'ensemble qu'on note classiquement $H_k(P_1,\dots,P_n)$, constitué des spécialisations $(x_1,x_2,\dots,x_r)$ des indéterminées $(X_1,\dots,X_r)$ pour lesquelles les polynômes $P_i(x_1,\dots,x_r,Y_1,\dots,Y_s)$, $i= 1,2,\dots,n$, sont irréductibles dans $k[Y_1,\dots,Y_s]$, contient beaucoup d'éléments de k^r. Précisément, on appelle partie hilbertienne de k^r tout ensemble, intersection d'un ensemble du type $H_k(P_1,\dots,P_n)$ avec un ouvert de Zariski de k^r et on dit que le corps k est hilbertien si pour tout entier $r\geq 1$, les parties hilbertiennes sont non vides. On appelle aussi ensemble mince tout ensemble dont le complémentaire contient une partie hilbertienne. En ces termes, le théorème d'irréductibilité de Hilbert s'énonce :

<u>Théorème 0</u> – Le corps $\mathbb{Q}$ des nombres rationnels est un corps hilbertien.

Le théorème de Hilbert autorise dans certaines situations à spécialiser des paramètres, sans modifier la structure algébrique. Ainsi, on peut spécialiser des indéterminées d'un polynôme de telle sorte que l'irréductibilité soit conservée, mais aussi la structure de son groupe de Galois. Cela permet de construire des extensions de $\mathbb{Q}$ de groupe de Galois donné G (problème inverse de la théorie de Galois) : il suffit de savoir construire une extension d'un corps d'indéterminées $\mathbb{Q}(X_1,\dots,X_r)$ de groupe G et de spécialiser ensuite $X_1,\dots,X_r$. On trouve dans les oeuvres de Hilbert, des exemples de ce genre de construction avec $G=S_n$ et $G=A_n$ (voir [27] pour d'autres exemples).

Le théorème de Hilbert sert également à construire des courbes elliptiques de rang élevé. La stratégie est la même : on construit une courbe elliptique sur un corps d'indéterminées qu'on spécialise ensuite. D'après un théorème de Néron, en dehors d'une ensemble mince, le rang se conserve. Néron a obtenu de cette façon des courbes elliptiques de rang sur $\mathbb{Q}$ r=9, 10 et même 11 [21]; sa construction pour r=11 a été précisée récemment par M. Fried [14]. Jusqu'aux travaux de J.-F. Mestre [20], c'était la seule méthode.

Pour montrer qu'un corps k est hilbertien, l'étude du cas $r=s=1$ (c'est-à-dire un paramètre et une indéterminée) est suffisante ([19] Ch. 9 § 3). Le problème consiste alors, via une réduction classique ([19] Ch. 9 Prop. 1.1), à compter des points sur des courbes algébriques, précisément à montrer que si $\deg_Y P_i \geq 2$ pour $i= 1,2,\ldots,n$, les ensembles

$$V'_k(P_1,\ldots,P_n)= \{x\in k/P_i(x,Y) \text{ n'a pas de racine dans } k \text{ pour } i=1,2,\ldots,n\}$$

sont infinis. Le théorème de Siegel sur la finitude des points entiers sur les courbes de genre ≥ 1 permet alors de majorer en $O(\sqrt{N})$ le nombre des entiers x qui ne sont pas dans $H_{\mathbb{Q}}(P_1,\ldots,P_n)$ et tels que $|x|\leq N$; cette estimation est d'ailleurs la meilleure possible à cause des carrés $(P_1= Y^2-X)$. On a des résultats semblables pour les x rationnels de hauteur $\leq N$ (cf. [27]).

Pour les dimensions plus grandes , des estimations analogues ont été établies par S.D. Cohen dans la fin des années 70 [5], mais la méthode est différente : elle consiste à étudier la réduction modulo un idéal premier $\mathfrak{p}$ d'un ensemble mince M; grâce au théorème de Lang-Weil sur le nombre de points mod $\mathfrak{p}$ d'une variété algébrique, on montre que certaines classes ne sont pas atteintes. Le théorème du grand crible permet d'en déduire que l'ensemble M lui-même n'a pas beaucoup d'éléments. La première partie de la méthode redonne d'autre part un résultat établi par A. Schinzel [24] et de façon effective par M. Fried [13] : toute partie hilbertienne de $\mathbb{Q}$ contient une progression arithmétique d'entiers. Pour plus de détails sur cet exposé introductif, nous renvoyons à [27] et [19].

Des travaux de ces dernières années permettent de donner une autre description des parties hilbertiennes, plus qualitative : ils mettent en évidence une relation liant la structure arithmétique d'un polynôme spécialisé $P(x,Y)$ (sa décomposition en polynômes irréductibles) à celle de x (sa décomposition en nombres premiers, ou plus généralement en idéaux premiers pour x algébrique). Cette relation se trouve être contraignante et imposera que "très souvent", on ne puisse être que dans le cas le plus simple, c'est-à-dire celui où $P(x,Y)$ est irréductible. On obtiendra de cette manière de nouveaux résultats liés au théorème d'irréductibilité de Hilbert et ce qui est important des résultats complètement explicites (voir en particulier le théorème 4).

<u>Notations</u>. Les valeurs absolues associées aux places d'un corps de nombres F sont normalisées de telle sorte qu'elles soient égales sur $\mathbb{Q}$ aux valeurs absolues usuelles. M_F désigne l'ensemble des places de F. La formule du produit s'écrit

$$\prod_{v\in M_F} |\xi|_v^{d_v^F} = 1 \quad \text{pour} \quad \xi \in F,\ \xi \neq 0$$

et la hauteur (logarithmique) d'un nombre algébrique est définie par

$$h(\xi)=\frac{1}{[F:\mathbb{Q}]}\sum_{v\in M_F} d_v^F \log\max(1,|\xi|_v) \quad \text{pour} \quad \xi \in F,$$

où d_v^F désigne le degré local de la place $v\in M_F$ par rapport à $\mathbb{Q}$. Enfin pour $\xi\in M_F$ on note $M_F(\xi)$ l'ensemble des places $v\in M_F$ où $|\xi|_v<1$ (par exemple, si ξ est un entier rationnel, $M_{\mathbb{Q}}(\xi)$ est l'ensemble des nombres premiers divisant ξ).

Dans la suite, k désigne un corps de nombres et P un polynôme irréductible dans $k(X)[Y]$. On suppose qu'il existe une série de Laurent $\underset{\sim}{Y} = \sum_{m\geq m_0} \eta_m X^m$ à coefficients η_m dans $\overline{\mathbb{Q}}$, solution de $P(X,\underset{\sim}{Y})=0$. D'un point de vue géométrique, cette hypothèse signifie que dans un modèle projectif lisse de la courbe "$P(x,y)=0$", la fonction x possède au moins un zéro simple Q.

Soit K le corps $K=k((\eta_m)_{m\geq m_0})$. Il est facile de voir que K est un corps de nombres et que $[K:k] \leq \deg_Y P$. Géométriquement K est le corps de définition sur k de Q.

Pour toute place v de K, notons R_v le rayon de convergence v-adique de $\underset{\sim}{Y}$ (on a $R_v>0$) et Y_v la fonction naturellement induite par $\underset{\sim}{Y}$ sur la boule ouverte épointée $B^*(0,R_v)= \{x\in K_v/0<|x|_v<R_v\}$ du complété K_v de K pour la métrique v.

L'énoncé suivant est fondamental. Pour ξ non nul dans k, on se donne a priori un diviseur Π dans $k[Y]$ du polynôme $P(\xi,Y)$; les nombres $Y_v(\xi)$ qui sont définis sont tous racines du polynôme $P(\xi,Y)$; dans le théorème 1, on s'intéresse seulement à ceux de ces nombres qui sont racines du polynôme Π.

<u>Théorème 1</u> - Soient $\xi\in k$, $\xi\neq 0$ et Π un diviseur de $P(\xi,Y)$ dans $k[Y]$. Soit $S(\xi,\Pi)$ l'ensemble des places v de K vérifiant :

$$|\xi|_v < R_v \quad \text{et} \quad \Pi(Y_v(\xi)) = 0.$$

On a alors

$$\frac{1}{[K:\mathbb{Q}]} \sum_{v\in S(\xi,\Pi)} d_v^K \log \min(1,|\xi|_v) = -\frac{\deg \Pi}{\deg_Y P} h(\xi) + O(\sqrt{h(\xi)})$$

où les constantes intervenant dans le $O(\ldots)$ ne dépendent que de P.

On peut interpréter le théorème 1 comme un résultat sur la distribution des $Y_v(\xi)$ qui sont définis, à l'intérieur des racines de $P(\xi,Y)$. En effet, à cause de la formule du produit, on a

$$h(\xi) = -\frac{1}{[K:\mathbb{Q}]} \sum_{v\in M_K} d_v^K \log \min(1,|\xi|_v);$$

la relation du théorème 1 signifie donc, qu'à un $O(1/\sqrt{h(\xi)})$ près, la probabilité (selon la loi image par l'application $v \longrightarrow d_v^K \log|\xi|_v$) qu'à une place v d'appartenir à un ensemble $S(\xi,\Pi)$, c'est-à-dire la probabilité qu'à un nombre $Y_v(\xi)$ d'être une racine de Π, vaut $\deg \Pi/\deg_Y P$.

Le théorème 1 est banal dans les cas extrêmes $\Pi=1$ et $\Pi = P(\xi,Y)$. Les situations intermédiaires sont plus intéressantes : la relation du théorème 1 impose à ξ des conditions (arithmétiques) non triviales pour qu'une telle situation puisse se produire c'est-à-dire pour que le polynôme $P(\xi,Y)$ puisse être réductible. Le corollaire suivant est une bonne illustration de ce lien existant entre les structures arithmétiques de $P(\xi,Y)$.

<u>Corollaire 1</u> ([10] § 2.3) - Soit $P(\xi,Y) = u\, P_1^{\alpha_1} \ldots P_g^{\alpha_g}$ la décomposition du polynôme $P(\xi,Y)$ en irréductibles distincts de $k[Y]$. Alors, pourvu que $h(\xi) > h_0$ où h_0 est une constante ne dépendant que de P, on a

$$g \leq \operatorname{Card} M_K(\xi).$$

Il suffit de remarquer que si $h(\xi)$ est suffisamment grand, le terme de gauche dans la relation du théorème 1 ne peut être nul si $\deg \Pi \geq 1$; par conséquent, on a $S(\xi,P_i) \cap M_K(\xi) \neq 0$ pour $i = 1,2,\ldots,g$. D'autre part, on a $S(\xi,P_i) \cap S(\xi,P_j) = \emptyset$ si $i \neq j$. Il y a donc au moins autant de places dans $M_K(\xi)$ que de polynômes P_i □

On obtient en particulier le résultat suivant :

<u>Corollaire 2</u> [29] - Soit P un polynôme irréductible dans $\mathbb{Q}(X)[Y]$. On suppose que le polynôme $P(0,Y)$ possède une racine simple dans $\mathbb{Q}$. ALors si p est un nombre premier et m un entier, le polynôme $P(p^m,Y)$ est irréductible dans $\mathbb{Q}[Y]$ dès que p^m est suffisamment grand (supérieur à une constante ne dépendant que de P).

En effet, on a ici $K=\mathbb{Q}$; $M_{\mathbb{Q}}(p^m)=\{p\}$; d'après le corollaire 1, on a donc $g=1$ dès que $h(p^m)= \log p^m$ est assez grand □ (voir [10] § 2.3 pour des généralisations du corollaire 2).

Le théorème 1 généralise les travaux antérieurs sur les valeurs de fonctions algébriques de T. Schneider [25] [26], P. Bundschuh [3] et V.G. Sprindzuk [29] [30] [31] [32]. Dans ces premiers travaux qui tous ont leur source dans le célèbre article de Siegel [28], seule une place entre en jeu (une place archimédienne chez Schneider et Bundschuh, une place finie chez Sprindzuk). En 1983, Sprindzuk donnera sa forme quasi définitive au résultat, en tenant compte simultanément de toutes les places [32]. Le théorème 1 [7] [10], obtenu grâce à une méthode différente de celle de Sprindzuk, affine un peu les hypothèses de son résultat et surtout en améliore les constantes : celles de Sprindzuk dépendent de k.

Les méthodes utilisées pour démontrer ce genre de résultats proviennent de la théorie des nombres transcendants.

La méthode de Siegel. C'est la méthode utilisée par Sprindzuk. Mise en oeuvre à l'origine par Siegel pour montrer la transcendance de valeurs de E-fonctions [28], on l'applique ici à des fonctions algébriques. Schématiquement, on procède de la façon suivante (voir [32] pour un exposé précis de la méthode).

Grâce au principe des tiroirs, on construit une fonction auxiliaire non nulle $\phi(X,\underset{\sim}{Y})$ où $\phi\in k[X,Y]$ est un polynôme dont on contrôle la hauteur, vérifiant les deux conditions :

(a) $\mathrm{ord}_0\, \phi(X,\underset{\sim}{Y}) \geq M$ où M est grand (précisément, M est un paramètre qu'on fait tendre vers $+\infty$ en fin de démonstration; ord_0 désigne la valuation X-adique sur $\overline{\mathbb{Q}}((X))$).

(b) Il existe $w\in S(\xi,\Pi)$ tel que $\phi(\xi,Y_w(\xi)) \neq 0$.
Par définition de $S(\xi,\Pi)$ on a aussi

(c) $\Pi(Y_w(\xi)) = 0$.

Enfin, on peut supposer Π irréductible : le cas général s'en déduit aisément. De (b) et (c) on déduit alors que R le résultant des polynômes Π et $\phi(\xi,Y)$ est non nul dans k. On applique la formule du produit à ce nombre algébrique :

$$\prod_{v\in M_K} |R|_v^{d_v^K} = 1.$$

Le résultat découle alors de la majoration de chacun des termes $|R|_v$. Ajoutons seulement que pour $v\in S(\xi,\Pi)$, on majore $|R|_v$ en

$|\phi(\xi, Y_v(\xi))|_v$ qui est petit à cause de (a) : il varie en $|\xi|_v^M$. Cela explique comment apparaît le terme de gauche dans le théorème 1. Le terme $\frac{\deg \Pi}{\deg_Y P} h(\xi)$ provient lui de la majoration de la hauteur de ϕ donnée par le principe des tiroirs □

Il y a cependant une difficulté. Pour obtenir la condition (b), on est obligé de faire des différentiations, ce qui fait apparaître des factorielles. Quand on travaille avec des E-fonctions, par exemple l'exponentielle, ces factorielles disparaissent, se simplifiant avec ceux qui figurent au dénominateur des coefficients du développement de Taylor des E-fonctions. En revanche, ils compliquent considérablement les estimations quand il s'agit de fonctions algébriques ou plus généralement de G-fonctions dont les coefficients de Taylor varient géométriquement. C'est d'ailleurs à cause de cette difficulté que les énoncés donnés par Siegel sur les G-fonctions resteront en suspens très longtemps : ce n'est qu'en 1981 qu'un résultat général sera démontré par Bombieri [1], au prix d'arguments très fins comme le théorème de Dwork-Robba. L'analogue de ce dernier résultat chez Sprindzuk est le lemme 4.5 de [32].

La méthode de Gel'fond. Il y a une alternative à la méthode de Siegel : la méthode de Gel'fond, que celui-ci élabora pour montrer la transcendance de a^b (septième problème de Hilbert). Adaptée au problème considéré, elle conduit d'une part au théorème 1 [6], d'autre part, dans le cadre plus général des G-fonctions, à une nouvel énoncé [10] tout à fait analogue à celui de Bombieri, et cela sans rencontrer l'écueil de la méthode précédente. Le principe de la démonstration est le suivant.

Grâce au principe des tiroirs, on construit une fonction auxiliaire non nulle $\phi(X,\underset{\sim}{Y})$ où $\phi \in k[X,Y]$ est un polynôme dont on contrôle la hauteur, vérifiant la condition
(a) pour toute place $v \in S(\xi,\Pi)$, la fonction $\phi(X,Y_v)$ a un zéro d'ordre élevé (supérieur à un paramètre M) au point ξ.
(Ici aussi, Π est supposé irréductible dans $k[Y]$ et la condition (a), en fait, ne dépend pas de $v \in S(\xi,\Pi)$, les valeurs des fonctions Y_v en ξ (ainsi que leur dérivées) étant conjuguées sur k pour $v \in S(\xi,\Pi)$).

Ensuite, on applique la formule du produit tout simplement au premier terme non nul γ du développement de $\phi(X,\underset{\sim}{Y})$ en 0 de $\phi(X,\underset{\sim}{Y})$, la condition (a) conduisant à des majorations de $|\gamma|_v$ en $|\xi|_v^M$ pour $v \in S(\xi,\Pi)$ □

Signalons pour clore ce chapitre que D.V et G.V Chudnovsky ont récemment obtenu, grâce à une méthode basée sur la théorie des approximants de Padé du second type, de nouveaux résultats sur la nature arithmétique des valeurs de G-fonctions $f_1,\ldots,f_n$ vérifiant des équations différentielles linéaires [4]. Leurs conclusions sont comparables à celles de [1] et [10], (à ceci près qu'ils ne travaillent qu'avec une seule place), mais au contraire de ces derniers travaux, ils se dispensent de toute hypothèse sur l'opérateur différentiel linéaire dont est supposé être solution le n-uple $(f_1,\dots,f_n)$; dans [1], on suppose que cet opérateur est fuchsien de type arithmétique, dans [10] que c'est un G-opérateur. Ils démontrent d'autre part ([4] th. III) que si, de plus, $f_1,\dots,f_n$ sont linéairement indépendantes sur $\overline{\mathbb{Q}}(X)$, alors le n-uple $(f_1,\dots,f_n)$ est nécessairement solution d'un opérateur différentiel vérifiant la condition de Galočkin [16], une troisième hypothèse classique dans ce genre de problème.

En 1983, Bombieri a donné une troisième démonstration du théorème 1 [2]. Sa nouvelle approche montre que le théorème 1, obtenu jusque là par des voies arithmétiques, a en fait une origine algébrique. La formulation de son résultat est plus géométrique, mais les énoncés sont équivalents (voir [8] Ch. 7).

Soit C une courbe projective irréductible lisse définie sur le corps de nombre k. Pour $Q\in C(k)$ un point k-rationnel sur C, on note λ_Q la fonction de Weil associée au diviseur Q (précisément, λ_Q désigne un représentant fixé de la classe, modulo les fonctions M_k-bornées sur C, des fonctions de Weil associées au diviseur Q (cf. [19] Ch. 10)). A v fixé, il faut voir $\lambda_Q(.,v)$ comme une valuation sur C : $\lambda_Q(M,v)$ grand signifie que "M est proche de Q pour la place v".

<u>Théorème 2</u>. - Soit φ une fonction rationnelle sur C définie sur k. Il existe une famille de nombres réels δ_v, $v\in M_k$, nuls pour toutes les places sauf un nombre fini ayant la propriété suivante : si $Q\in C(k)$ est un pôle de φ k-rationnel, alors pour tout point M dans $C(k)$, on a

$$\frac{1}{[k:\mathbb{Q}]}\sum_{\substack{v\in M_k\\ \lambda_Q(M,v)>\delta_v}} d_v^k \log\max(1,|\varphi(M)|_v) = -\frac{\mathrm{ord}_Q\varphi}{\deg\varphi}\, h(\varphi(M)) + O(\sqrt{h(\varphi(M))})$$

où les constantes intervenant dans le $O(\dots)$ dépendent de C et φ seulement.

La démonstration de Bombieri s'appuie sur deux résultats fondamentaux : le théorème de décomposition de Weil ([33], [19] Ch. 10) et la quadraticité de la hauteur sur les variétés abéliennes ([22], [19] Ch. 5). Son principe est le suivant (cf. [2] [9]).

En utilisant les propriétés standards des fonctions de Weil, notamment le théorème de décomposition de Weil, on montre l'existence d'une famille de nombres réels δ_v, $v\in M_k$, nuls pour presque tout v, telle que si $Q\in C(k)$ est un pôle de φ, alors pour tout $M\in C(k)$, on ait

$$\frac{1}{[k:\mathbb{Q}]} \sum_{\substack{v \in M_k \\ \lambda_Q(M,v) > \delta_v}} d_v^k \log \max(1, |\varphi(M)|_v) = -\operatorname{ord}_Q \varphi \; h_Q(M) + O(1)$$

où h_Q désigne la hauteur associée à la classe du diviseur Q dans le groupe de Picard de C.

Ensuite, on fait varier Q parmi les pôles de φ et on somme les égalités ainsi obtenues : cela donne

$$h(\varphi(M)) = \sum_{Q' \text{pôle de } \varphi} (-\operatorname{ord}_{Q'} \varphi) h_{Q'}(M) + O(1).$$

Mais à cause de la quadraticité de la hauteur, on a (cf. [19] Ch. 5 § 5)

$$h_{Q'} = h_Q + O(\sqrt{h_Q}) \qquad \text{pour tout } Q'.$$

L'égalité précédente donne alors

$$h(\varphi(M)) = \deg \varphi \; h_Q(M) + O(\sqrt{h(\varphi(M))})$$

ce qui, reporté dans la première égalité, fournit le résultat désiré □

Comme pour le théorème 1, notons que le terme de gauche dans la relation du théorème 2 ne peut être nul si la hauteur de $\varphi(M)$ est suffisamment grande. Supposons tous les pôles de φ rationnels sur k. La famille $(\delta_v)_{v \in M_k}$ du théorème 2 peut être choisie de telle sorte que les ensembles $\{v \in M_k / \lambda_Q(M,v) > \delta_v\}$ où Q varie dans l'ensemble des pôles de φ soient disjoints deux à deux. De ces deux remarques, on déduit que, dès que $h(\varphi(M))$ est assez grand, le nombre de places v de k où $|\varphi(M)|_v > 1$ est minoré par le nombre de pôles de φ. Dans le cas général, on peut faire le raisonnement précédent sur une extension de k et redescendre sur k grâce à des arguments galoisiens (cf. [10] § 2.4). On obtient le résultat suivant :

<u>Corollaire</u> - Soient φ une fonction rationnelle sur C définie sur k et μ le nombre de pôles de φ non conjugués sur k. Pour tout point M k-rationnel sur C, on a

$$\operatorname{Card} M_k(1/\varphi(M)) \geq \mu$$

dès que $h(\varphi(M))$ est assez grand (supérieur à une constante ne dépendant que de C et φ).

Enoncé sous des formes diverses, ce résultat se situe à la croisée de plusieurs travaux : [2] Th. p. 305, [10] § 2.4 Corollaire, [15] Prop. 4.4, [34]. Aux méthodes utilisées chez les trois premiers qui sont celles décrites dans cet exposé, R. Weissauer [34] en ajoute une troisième qui s'appuie sur des arguments non-standards.

On peut, comme Bombieri [2], voir le corollaire comme un résultat de finitude des points entiers (ou S-entiers) sur certaines courbes algébriques. En particulier, il donne l'exemple suivant :

Soit P un polynôme irréductible dans $\mathbb{Q}[X,Y]$. Supposons que la partie homogène de plus haut degré dans P ne soit pas la puissance d'un irréductible de $\mathbb{Q}[X,Y]$. Géométriquement, cela impose que sur C, un modèle projectif lisse du corps de fonctions $\mathrm{Fract}\,(\overline{\mathbb{Q}}[X,Y]/P)$, la fonction x ait au moins deux pôles non conjugués sur $\mathbb{Q}$ (soit, avec les notations du corollaire, $\mu \geq 2$ pour $\varphi=x$). On déduit donc du corollaire qu'il n'y a qu'un nombre fini de points M $\mathbb{Q}$-rationnels sur C vérifiant $\mathrm{Card}(M_{\mathbb{Q}}(1/x(M))<2$. En particulier les points M, $\mathbb{Q}$-rationnels sur C pour lesquels $x(M)\in\mathbb{Z}$ sont en nombre fini; l'équation $P(x,y)=0$ n'admet donc qu'un nombre fini de solutions entières x,y. On retrouve ici un résultat de Runge [23].

Il est intéressant de noter que le théorème 2 et son corollaire restent valides si le corps k est plus généralement un corps muni d'un ensemble de valeurs absolues satisfaisant la formule du produit ([19] Ch. 2). En effet, on peut déduire directement du corollaire que le corps k est hilbertien. On obtient ainsi que tout corps avec une formule du produit est un corps hilbertien, un résultat dû à Weissauer.

Voici comment on procède pour déduire du corollaire l'hilbertianité de k. Les idées suivantes sont de M. Fried ([15] th. 4.2). Cependant, contrairement à lui, nous n'utiliserons pas ici l'existence d'un ultrafiltre maximal non trivial sur $\mathbb{N}$.

Via les réductions classiques rappelées en introduction, il s'agit de démontrer que si $P_1,\dots,P_n\in k(X)[Y]$ sont n polynômes absolument irréductibles de degré ≥ 2, alors l'ensemble $V_k'(P_1,\dots,P_n)$ est infini.

Remarquons tout d'abord que, quitte à changer X en $b+1/X$, pour b convenablement choisi dans k, on peut supposer que les corps de rupture sur $k(X)$ des polynômes P_i, $i=1,2,\dots,n$, ne sont pas ramifiés au-dessus de la place $x=\infty$.

Soit H une partie infinie de $k-\{0\}$ ayant la propriété suivante :

(1) Il existe un entier ℓ tel que pour tout x dans H, $\mathrm{Card}\, M_k(x) \leq \ell$.

Nous allons montrer que

(2) Il existe $a\in k$ tel que l'ensemble $a+1/H\cap V_k'(P_1,\dots,P_n)$ soit infini, ce qui, en notant $P_i^{(a)}=P_i(a+1/X,Y)$ pour $i=1,2,\dots,n$ et $a\in k$, équivaut à

(2') Il existe $a\in k$ tel que l'ensemble $H\cap V'_k(P_1^{(a)},\dots,P_n^{(a)})$ soit infini.

Pour m un entier quelconque, on commence par construire par récurrence $a_1,\dots,a_m$ dans k de telle sorte que pour $j=2,3,\dots,m$, les discriminants des polynômes $P_i^{(a_j)}$, $i=1,2,\dots,n$ d'une part, et les discriminants des polynômes $P_i^{(a_\nu)}$, $i=1,2,\dots,n$, $\nu=1,\dots,j-1$, d'autre part, n'aient pas de racines en commun : cela est clairement réalisable vu la forme des polynômes $P_i^{(a)}$.

On raisonne ensuite par l'absurde. Notons $V_k(P_i^{(a)})$ l'ensemble des éléments x de k tels que le polynôme $P_i^{(a)}(x,Y)$ ait une racine dans k; si (2') est faux, l'ensemble $H\cap\bigcup_{j=1}^{m} V'_k(P_1^{(a_j)},\dots,P_n^{(a_j)})$ est fini. On en déduit que l'ensemble
$H\cap\bigcup_{1\leq i_1,\dots,i_m\leq n} V_k(P_{i_1}^{(a_1)})\cap\dots\cap V_k(P_{i_m}^{(a_m)})$ est infini, et donc qu'il existe un m-uplet $(i_1,\dots,i_m)$ dans $\{1,\dots,n\}$ pour lequel

(3) l'ensemble $H\cap V_k(P_{i_1}^{(a_1)})\cap\dots\cap V_k(P_{i_m}^{(a_m)})$ est infini.

Notons pour $j=1,2,\dots,m$, y_j un zéro dans $\overline{k(X)}$ du polynôme $P_{i_j}^{(a_j)}$ et L le corps $L=k(X,y_1,\dots,y_m)$. A cause du choix des nombres $a_1,\dots,a_m$ pour $j=2,\dots,m$, l'intersection des clôtures normales sur $k(X)$ des extensions $k(X,y_j)$ et $k(X,y_1,\dots,y_{j-1})$ est ramifiée nulle part, donc vaut $k(X)$ (équivalent classique sur $k(X)$ du théorème de Hermite-Minkowsky). Ces extensions sont donc linéairement disjointes sur $k(X)$. En particulier, on a

(4) $$[L : k(X)] = \deg P_{i_1}\dots\deg P_{i_m}.$$

Considérons maintenant C (resp. C_j) un modèle projectif lisse du corps de fonctions $L\overline{k}$ (resp. $\overline{k}(X,y_j)$). (3) signifie qu'il existe une infinité de points M k-rationnels sur C vérifiant $x(M)\in H$. On déduit alors de la définition de H (1) et du corollaire (qu'on applique à $\varphi=1/x$), qui si μ désigne le nombre de zéros non conjugués sur k de la fonction x, alors

$$\mu \leq \ell .$$

Soit Q l'un de ces zéros. Il s'envoie par restriction (en le voyant comme une place du corps $L\overline{k}$ au dessus de $X=0$), sur un zéro Q_j de la fonction x sur la courbe C_j. Ce dernier, par la transformation $X \longrightarrow a_j+1/X$, correspond à un pôle de la fonction x sur la courbe "$P_{i_j}(x,y)=0$", c'est-à-dire l'une des courbes "$P_1(x,y)=0$",...,"$P_n(x,y)=0$". Conclusion : le corps de définition sur k du point Q qui est isomorphe au compositum des corps de définition sur k des points Q_j, $j=1,2,\ldots,m$ (à cause de la condition sur $a_1,\ldots,a_m$), a un degré sur k qui peut être majoré par un nombre r indépendant de m. Enfin à cause de la réduction faite en début de démonstration, on a $\mathrm{ord}_Q x=1$. Toutes ces remarques conduisent finalement à

$$\deg \varphi = \sum_{\substack{Q \\ x(Q)=0}} \mathrm{ord}_Q\, \varphi \leq l_r .$$

Il suffit, pour obtenir la contradiction désirée, de choisir m assez grand, puisque, à cause de (4), on a aussi

$$\deg \varphi = [L : k(X)] \geq 2^m \quad \square$$

La ramification est l'un des outils de base de la démonstration précédente. Elle joue également un rôle primordial dans un travail récent de R. Dvornicich et U. Zannier [12]. Soit P un polynôme irréductible dans $\mathbb{Q}[X,Y]$; pour $a\in\mathbb{Q}$, notons $\theta(X+a)$ une fonction algébrique solution de $P(X+a,\theta(X+a))=0$. En reprenant l'argument développé dans la démonstration précédente, on construit facilement $a_1,\ldots,a_m$ dans $\mathbb{Q}$ tels que l'extension $\mathbb{Q}(X,\theta(X+a_1),\ldots,\theta(X+a_m))$ soit de degré maximal sur $\mathbb{Q}(X)$, c'est-à-dire $(\deg_Y P)^m$.

Dvornicich et Zannier étudient l'analogue de ce problème, pour des valeurs de fonctions algébriques : si, pour $m\in\mathbb{N}$, on note θ_m une racine dans $\overline{\mathbb{Q}}$ du polynôme $P(m,Y)$, que peut-on dire du degré sur $\mathbb{Q}$ du corps $K(m)=\mathbb{Q}(\theta_1,\ldots,\theta_m)$.

Pour $m\in\mathbb{N}$, définissons l'entier $D(m)$ par

$$D(m)= \min_{\theta_1,\ldots,\theta_m} [\mathbb{Q}(\theta_1,\ldots,\theta_m):\mathbb{Q}]$$

où le "min" porte sur l'ensemble des m-uplets $(\theta_1,\ldots,\theta_m)$ vérifiant $P(i,\theta_i)=0$, $i=1,\ldots,m$. L'exemple du polynôme $P=Y^q-X$ pour lequel $K(m)=\mathbb{Q}(\sqrt[q]{p}$, p premier, $p\leq m)$ et donc $D(m)\leq q^{\Pi(m)}$ avec $\Pi(m)\simeq m/\log m$, montre qu'on ne peut pas espérer une croissance géométrique de $D(m)$. En fait, cet exemple est significatif puisque Dvornicich et Zannier démontrent que, pour P quelconque, il existe une constante $C>1$ vérifiant $D(m) \geq C^{m/\log m}$ pour m assez grand ([12] th. 2).

le résultat provient d'une étude de la ramification des corps $\mathbb{Q}(\theta_m)$, l'idée directrice étant de montrer que pour "beaucoup" d'entiers m, il existe un nombre premier p ramifié dans $\mathbb{Q}(\theta_m)$ et pas dans $\mathbb{Q}(\theta_j)$, $j= 1,2,\ldots,m-1$ de telle sorte qu'on puisse conclure que $K(m) \supsetneq K(m-1)$ pour "beaucoup" d'entiers m.

Le résultat-clé relie la ramification des corps $\mathbb{Q}(\theta_m)$ à celle du corps $\mathbb{Q}(X,\theta(X))$, de façon précise, Δ désignant le discriminant de l'extension $\mathbb{Q}(X,\theta(X))$ de $\mathbb{Q}(X)$, les nombres premiers p qui se ramifient dans $\mathbb{Q}(\theta_m)$ à ceux qui divisent $\Delta(m)$ (cf. [12] lemmes 3 et 4 pour des énoncés précis). On conclut grâce à des résultats classiques qui permettent d'estimer le nombre de premiers p pour lesquels l'équation $\Delta(m)=0$ a une solution dans $\mathbb{F}_p$.

Terminons cet exposé par une application spectaculaire des théorèmes 1 et 2. On appelle partie hilbertienne universelle (de $\mathbb{Q}$) toute suite $(x_m)_{m\geq 0}$ de nombres rationnels x_m ayant la propriété suivante : pour tout polynôme P irréductible dans $\mathbb{Q}(X)[Y]$, le polynôme $P(x_m,Y)$ est irréductible dans $\mathbb{Q}[Y]$ dès que $m \geq m(P)$ où $m(P)$ est une constante ne dépendant que de P. En d'autres termes, c'est une partie infinie de $\mathbb{Q}$ incluse dans toute partie hilbertienne de $\mathbb{Q}$, à un ensemble fini près. Commençons par quelques remarques :

(a) Pour e un entier quelconque supérieur à 2, il ne peut y avoir dans une partie hilbertienne universelle qu'un nombre fini de termes qui sont des puissances e-ièmes dans $\mathbb{Q}$ (considérer $P= Y^e-X$). En particulier, la suite des entiers positifs, la suite $(t^m)_{m\geq 0}$ où t est un rationnel fixé, ne sont pas des parties hilbertiennes universelles.

(b) La suite $(p_m)_{m\geq 0}$ des nombres premiers n'est certainement pas une partie hilbertienne universelle. En effet, on conjecture classiquement qu'il existe une infinité de nombres premiers p de la forme $p= y^2+1$ avec y entier, ce qui signifie que pour $P= Y^2+1-X$, le polynôme $P(p,Y)$ se décompose pour une infinité de nombres premiers p.

(c) L'image d'une suite $(x_m)_{m\geq 0}$, qui n'est pas une partie hilbertienne universelle, par une homographie rationnelle bijective $x \longrightarrow \frac{ax+b}{cx+d}$ n'est pas non plus une partie hilbertienne universelle.

L'existence de parties hilbertiennes universelles est une conséquence du théorème d'irréductibilité de Hilbert : on peut ordonner en une suite $(P_n)_{n\geq 0}$ l'ensemble des polynômes irréductibles dans $\mathbb{Q}(X)[Y]$; d'après le théorème de Hilbert, pour tout entier $m\geq 0$, l'ensemble $H_{\mathbb{Q}}(P_0,P_1,\dots,P_m)$ est non vide; on choisit x_m dedans. Alors la suite $(x_m)_{m\geq 0}$ est une partie hilbertienne universelle.

Mais jusqu'en 1981, on ne disposait d'aucun exemple explicite de partie hilbertienne universelle. On peut désormais en construire grâce aux théorèmes 1 et 2. Le premier a été donné par Sprindzuk [31]. Il s'agit de la suite

$$x_m = [\exp(\sqrt{\log(\log m)})] + m!2^{m^2}, \qquad m\geq 3.$$

Nous allons donner ici un deuxième exemple. Sa construction repose sur le résultat suivant [11].

<u>Théorème 3</u> - Soient $P\in\mathbb{Q}[X,Y]$ un polynôme de degré partiel en Y supérieur ou égal à 2 et $e\geq 1$ un entier tels que le polynôme $P(X^e,Y)$ soit absolument irréductible et possède une racine $\underset{\sim}{Y}$ dans $\overline{\mathbb{Q}}((X))$. Soit b un entier distinct de $0,1,-1$. Il existe un entier non nul $\alpha_0(P,b)$ tel que pour tout multiple non nul α de $\alpha_0(P,b)$, le polynôme $P(\alpha^e b^m,Y)$ n'ait pas de racine dans $\mathbb{Q}$ si m est un entier assez grand (précisément supérieur à une constante $m_0(P,b,\alpha)$ ne dépendant que de P,b,α).

Supposons pour simplifier $e=1$ (le cas général, un peu plus délicat, est traité dans [11]). On est alors sous les hypothèses du théorème 1. Rappelons que K désigne le corps engendré sur $\mathbb{Q}$ par les coefficients de $\underset{\sim}{Y}$. Pour démontrer le théorème 3, on distingue deux cas.

<u>Premier cas</u> : $[K:\mathbb{Q}] < \deg_Y P$ - On choisit $\alpha_0=1$. Soient α un entier non nul et Π un diviseur dans $\mathbb{Q}[Y]$ du polynôme $P(\alpha b^m,Y)$ de degré $\deg \Pi \geq 1$. Notre objectif est de montrer que $\deg \Pi \geq 2$.

Notons S_m l'ensemble $S_m = S(\alpha b^m,\Pi) \cap M_K(b)$. Le théorème 1 donne

$$\frac{1}{[K:\mathbb{Q}]}\sum_{v\in S_m} d_v^K \log|b|_v + \frac{\deg \Pi}{\deg_Y P}\, h(b) = O\left(\sqrt{\frac{h(b)}{m}}\right).$$

Notons x_m le terme de gauche. Comme $1 \leq \deg \Pi \leq \deg_Y P$ et que $S_m \subset M_K(b)$, la suite $(x_m)_{m\geq 1}$ ne prend qu'un nombre fini de valeurs. Comme, d'après la relation précédente, elle tend vers 0, elle est nulle à partir d'un certain rang.

b est un nombre rationnel; la relation $x_m=0$ s'écrit donc :

$$\sum_{w\in M_{\mathbb{Q}}(b)} \left[\frac{1}{[K:\mathbb{Q}]} \sum_{\substack{v\in S_m \\ v/w}} d_v^K - \frac{\deg \Pi}{\deg_Y P}\right] \log |b|_w = 0.$$

Mais les nombres $\log |b|_w$, où w décrit l'ensemble $M_{\mathbb{Q}}(b)$ sont linéairement indépendants sur $\mathbb{Q}$. On obtient donc, que pour tout entier m assez grand, on a

$$[K : \mathbb{Q}] \deg \Pi = \deg_Y P \sum_{\substack{v\in S_m \\ v/w}} d_v^K \quad \text{pour tout } w\in M_{\mathbb{Q}}(b)$$

ce qui donne en particulier

$$\deg \Pi \geq \frac{\deg_Y P}{[K:\mathbb{Q}]} > 1.$$

<u>Deuxième cas</u> : $[K : \mathbb{Q}]= \deg_Y P$ - Moyennant un petit changement sur P, cette hypothèse signifie que le polynôme $P(0,Y)$ est irréductible dans $\mathbb{Q}[Y]$. D'après un résultat classique de Hasse, il existe un nombre premier p tel que l'équation $P(0,y)=0$ n'ait pas de solutions y dans $\mathbb{F}_p$. Prenons $\alpha_0=p$. Si α est un multiple de α_0, le polynôme $P(\alpha b^m,Y)$ ne peut avoir de racine rationnelle y puisqu'en passant aux classes modulo p, on aurait alors $P(0,y)\equiv 0$ modulo p □

Appliqué à plusieurs polynômes à la fois, le théorème 3 conduit à une nouvelle version du théorème d'irréductibilité de Hilbert [11] : si b est un entier distinct de 0,1 et -1, toute partie hilbertienne de $\mathbb{Q}$ contient une progression géométrique $(ab^m)_{m\geq 1}$ de raison b. Mais le théorème 3 permet d'aller plus loin encore.

Notons, pour tout nombre réel x, $p(x)$ et $\theta(x)$ les entiers définis par

$$\begin{cases} p(x)= \operatorname{Max} \{p/p \text{ premier}, \ p\leq x\} & \text{si } x\geq 2 \\ p(x)= 1 & \text{si } x<2 \end{cases}$$

$$\theta(x)= \prod_{\substack{p \text{ premier} \\ p\leq x}} p.$$

<u>Théorème 4</u> - Soit b un entier distinct de 0,1 et -1. Pour $m\geq 2$, soit

$$x_m= p(\log \log m)\, \theta(\log m)^{[\log \log m]!}\, b^m.$$

La suite $(x_m)_{m\geq 2}$ est une partie hilbertienne universelle. Autrement dit, si P est un polynôme irréductible dans $\mathbb{Q}(X)[Y]$, alors dès que m est suffisamment grand, le polynôme $P(x_m,Y)$ est irréductible dans $\mathbb{Q}[Y]$.

Le gros du travail restant à faire consiste à préciser le théorème 3. En utilisant un résultat effectif de J.C. Lagarias, H.L. Montgomery et A.M. Odlyzko sur le théorème de Chebotarev [18], dont le lemme de Hasse est un corollaire, on montre qu'on peut choisir pour $\alpha_0(P,b)$ un nombre premier vérifiant

$$\alpha_0(P,b) \leq \nu(P,b)$$

où $\nu(P,b)$ est une constante qu'on peut calculer effectivement, ne dépendant que de P,b. Le calcul des constantes intervenant dans le théorème 1, qui est fait dans [10], fournit d'autre part une majoration explicite de $m_0(P,b,\alpha)$ en fonction de P, b et de α.

On procède alors de la façon suivante. Il s'agit de montrer que si $P\in\mathbb{Q}[X,Y]$ est un polynôme absolument irréductible de degré partiel en Y supérieur ou égal à 2, le polynôme $P(x_m,Y)$ n'a pas de racine dans $\mathbb{Q}$ si m est un entier assez grand.

Soit $e\geq 1$ un entier tel que le polynôme $P(X^e,Y)$ admette une racine dans $\overline{\mathbb{Q}}((X))$; l'existence de e est assurée par le théorème de Puiseux. On introduit ensuite le polynôme $P_m = P(p(\log\log m)X,Y)$, défini pour $m\geq 2$. Enfin, notons α_m le nombre $\alpha_m = \theta(\log m)^{[\log\log m]!/e}$; c'est un entier dès que m est assez grand. Deux cas se présentent.

<u>Premier cas</u> : $P(X^e,Y)$ n'est pas absolument irréductible - Le polynôme $P_m(X^e,Y)$ est alors lui aussi non absolument irréductible. Par contre, d'après la proposition 3 du paragraphe 4 de [10], dès que m est assez grand, il est irréductible dans $\mathbb{Q}(\beta)[X,Y]$, où β désigne une racine e-ième de b. Sous ces conditions, le polynôme $P_m((\alpha_m\beta^m)^e,Y) = P(x_m,Y)$ ne peut avoir de racine dans $\mathbb{Q}(\beta)$ (en particulier dans $\mathbb{Q}$) que si elle est multiple. Cela ne peut se produire que pour un nombre fini d'entiers m.

<u>Deuxième cas</u> : $P(X^e,Y)$ est absolument irréductible - Le polynôme P_m vérifie alors les hypothèses du théorème 3. Quelques calculs basés sur les estimations préliminaires des constantes $\alpha_0(P,b)$ et $m_0(P,b,\alpha)$ montrent que, dès que m est assez grand

$\log m \geq \nu(P_m,b)$ de telle sorte que $\alpha_0(P_m,b)$ divise α_m et $m \geq m_0(P_m,b,\alpha_m)$.

Pour m assez grand, le polynôme $P_m(\alpha_m^e b^m,Y) = P(x_m,Y)$ n'a donc pas de racine dans $\mathbb{Q}$. $\square$

BIBLIOGRAPHY

[1] E. Bombieri.- On G-functions, Recent progress in analytic number theory, H. Halberstam and C. Hooley ed., Acad. Press (1981), vol. 2, 1-67.

[2] E. Bombieri.- On Weil's "Théorème de Décomposition", Amer. J. Math., 105 (1983), 295-308.

[3] P. Bundschuh.- Une nouvelle application de la méthode de Gel'fond. Sem. Delange-Pisot-Poitou, Théorie des Nombres, 19ème année (1977-78), N° 42.

[4] D.V. and G.V. Chudnovsky.- Applications of Padé approximations to diophantine inequalities in values of G-functions, Number Theory, Sem. N.Y., 1983-84, Lect. Notes Math. 1135, (1985), 9-51.

[5] S.D. Cohen.- The distribution of the Galois groups of integral polynomials, Illinois J. Math. (1979), Vol. 23, N°1, 135-152.

[6] P. Dèbes.- Une version effective du théorème d'irréductibilité de Hilbert, Sém. Anal; Ultramétrique, Amice-Christol-Robba, 10ème année (1982-83), N° 10.

[7] P. Dèbes.- Spécialisations de polynômes, Math. rep. Acad. Sc., Royal Soc. Canada, Vol. V, n° 6, (Dec. 1983).

[8] P. Dèbes.- Valeurs algébriques de fonctions algébriques et théorème d'irréductibilité de Hilbert, Thèse 3ème cycle, Univ. P. et M. Curie (Paris VI), (1984).

[9] P. Dèbes.- Quelques remarques sur un article de Bombieri concernant le théorème de décomposition de Weil, Amer. J. Math. 107 (1985), 39-44.

[10] P. Dèbes.- G-fonctions et théorème d'irréductibilité de Hilbert, Acta Arithmetica, Vol. 47, N° 4, (à paraître).

[11] P. Dèbes.- Parties hilbertiennes et progressions géométriques, C.R. Acad. Sc. Paris, t. 302, Série I, n° 3, (1986).

[12] R. Dvornicich and U. Zannier.- Fields containing values of algebraic functions, Publ. Univ. Pisa (Novembre 1983).

[13] M. Fried.- On Hilbert's irreducibility theorem, J. Number Theory, 6 (1974), 211-231.

[14] M. Fried.- Constructions arising from Neron's high rank curves, Trans. Amer. Math. Soc. Vol. 281, N°2, 1984.

[15] M. Fried.- On the Sprindzuk-Weissauer approach to universal Hilbert subsets, Israel. J. Math. vol. 51, N° 4, 1985.

[16] A.L. Galochkin.- Lower bounds of polynomials in the values of a certain class of analytic functions, Math. Sb. 95 (1974), 396-417.

[17] D. Hilbert.- Uber die Irreduzibilität ganzer rationaler Funktionen mit ganzahligen Koeffizienten, Gesammelte Afhandlungen, Springer-Verlag (1983) [réimpression Chelsea (1965)] Vol. 2, N° 18, 264-286. Ou J. für die reine und angew. Math. 110 (1982), 104-129.

[18] J.C. Lagarias, H.L. Montgomery and A.M. Odlyzko.- The bound for the least prime ideal in the Chebotarev density theorem, Invent. Math. 54 (1979), 271-296.

[19] S. Lang.- Fundamentals of Diophantine Geometry, Springer-Verlag (1983).

[20] J.F. Mestre.- C.R. Acad. Sci. Paris 295 (1982), 643-644.

[21] A. Néron.- Problèmes arithmétiques et géométriques rattachés à la notion de rang d'une courbe algébrique dans un corps, Bull. Soc. Math. France, 80 (1952), 101-166.

[22] A. Néron.- Quasi-fonctions et hauteurs sur les variétés abéliennes, Ann. of Math. 82 (1965) n° 2, 249-331.

[23] C. Runge.- Veber ganzzahlige Lösungen von Gleichungen zwischen zwei Veränderlichen, J. für die reine und angew. Math. 100 (1887), 425-435.

[24] A. Schinzel.- On Hilbert's irreducibility theorem, Acta Arithmetica 16 (1965), 334-340.

[25] T. Schneider.- Rationale Punkte Uber ciner algebraischen Kurve, Sem. Delange-Pisot-Poitou, Théorie des Nombres, 15ème année (1973/74), N° 20.

[26] T. Schneider.- Eine bemerkung zu einem Satz von C.L. Seigel, Comm. pure and applied Math. 29 (1976), 775-782.

[27] J.-P. Serre.- Autour du théorème de Mordell-Weil, II, Cours au Collège de France, (1980/81), Notes rédigées par M. Waldschmidt.

[28] C.L. Siegel.- Uber Einige Anwandungen diophantischer Approximationen, Gesammelte Afhandlungen, Springer-Verlag (1966), vol. 1, N° 16, 209-266. Ou Abh. Preus. Akad. Wiss. Phys. Math. Kl. 1 (1929), 14-67.

[29] V.G. Sprindzuk.- Hilbert's irreducibility theorem and rational points on algebraic curves, Doklady Acad. Nauk. SSSR 247 (1979), 285-289.

[30] V.G. Sprindzuk.- Reducibility of polynomials and rational points on algebraic curves Doklady Acad. Nauk. SSSR 250 (1980), 1327-1330.

[31] V.G. Sprindzuk.- Diophantine equations involving unknown primes, Trudy M.I.A.N. SSSR 158 (1981), 180-186.

[32] V.G. Sprindzuk.- Arithmetic specializations in polynomials, J. Reine und Angew. Math. 340 (1983), 26-52.

[33] A. Weil.- Arithmetic on algebraic varieties, Annals of Math. 53 (1951), 412-444.

[34] R. Weissauer.- Hilbertsche Körper, Thesis, Heidelberg (1980).

Pierre DEBES
Department of Mathematics
University of Florida
Gainesville FLA 32611
U.S.A.

ELLIPTIC CURVES AND SOLUTIONS OF A-B=C

Gerhard FREY

Introduction.

In the following we want to report how elliptic curves can be used to get information about solutions of the equation A-B=C. One associates to A and B the curves

$$E_{(A,B)} : Y^2 = X(X-A)(X-B)$$

whose arithmetic reflects arithmetical properties of A,B and C.

In this paper we restrict ourselves to the case that A,B and C are rational integers; the general number field case can be found in [Fr 1].

We are mainly interested in the "Fermat type" case, to be more precise we state the following wellknown conjectures :

Asymptotic Fermat conjecture : Assume that for fixed natural relatively prime numbers a_1, a_2, a_3 the equation

$$a_1 Z_1^p - a_2 Z_2^p = a_3 Z_3^p$$

has solutions in $\mathbb{Z}\setminus\{0\}$ for infinitely many primes p. Then $a_1 \pm a_2 = \pm a_3$.

Fermat's conjecture : For $p \geq 3$ $Z_1^p - Z_2^p = Z_3^p$ has no nontrivial solution.

Our aim is to relate these conjectures with conjectures about elliptic curves.

At first we compare the height of E in the sense of Faltings (cf. [De]) or, more down to earth, the height of j_E of elliptic curves $E/\mathbb{Q}$ with the conductor of E, we state a height conjecture ($\underline{H}$) (which is true for elliptic curves over function fields) and which implies Szpiro's conjecture (cf. [Fr 2]). We show that ($\underline{H}$) is equivalent with the ($\underline{ABC}$)-conjecture of Masser-Oesterlé saying that for A, B and C= A-B with gcd(A,B)=1 one has :

$|A| \leq c(\epsilon)(\prod_{l|ABC} l)^{(1+\epsilon)}$ (Prop. 1). As a consequence we get that the asymptotic Fermat conjecture follows from ($\underline{H}$).

Next we discuss properties of points of order p of $E_{(A,B)}$ for $A= a_1 z_1^p$, $B= a_2 z_2^p$ and $A-B= a_3 z_3^p$ (Prop. 3) which are rather exotic for $a_i=1$ so that one suspects that in this case $E_{(A,B)}$ cannot exist.

In the second section we assume that $E_{(A,B)}$ is a modular elliptic curve, i.e. we assume that the conjecture ($\underline{TS}$) of Taniyama and Shimura holds, and then we determine other additional properties which stable elliptic curves have to have in order to satisfy ($\underline{H}$). Since $h(E)$ is linked in a natural way with the degree of modular parametrisations $\varphi : X_o(N) \longrightarrow E$ one is led to state the degree conjecture ($\underline{D}$) for φ which implies ($\underline{H}$) (and hence the asymptotic Fermat conjecture) (Prop. 4). The properties of points of order p of $E_{(A,B)}$ mentioned above lead to very special representations of the absolute Galois group $G_{\mathbb{Q}}$ of $\mathbb{Q}$ into $Gl(2,\mathbb{Z}/p)$ which should be linked with cusp forms according to a conjecture of Serre (cf. [Se 2], [Se 4]), a rather weak form is the level conjecture ($\tilde{\underline{S}}$). In proposition 5 we show that ($\underline{TS}$)+($\tilde{\underline{S}}$) implies Fermat conjecture as well as results closely related with the asymptotic Fermat conjecture. Recently Ribet succeeded to prove that ($\tilde{\underline{S}}$) is true in the prime level case (cf. [Ri 2]) and it seems that he can prove ($\tilde{\underline{S}}$) for arbitrary squarefree level too ! Since $E_{(A,B)}$ is stable its conductor is squarefree, and hence Ribet's result implies : The Taniyama-Shimura conjecture implies Fermat's conjecture.

§1. The curves $E_{(A,B)}$.

For $A, B \in \mathbb{Z}$ with $\gcd(A,B)=1$, A even and $B \equiv 1 \bmod 4$ define the elliptic curve

$$E_{(A,B)} : Y^2 = X(X-A)(X-B).$$

This curve has stable reduction in all primes $l \neq 2$, i.e. either E has good reduction or reduction of multiplicative type mod l. If $2^4 || A$ $E_{(A,B)}$ has good reduction mod 2, and if $2^5 | A$ it has reduction of multiplicative type at 2. Its absolute invariant is

$$j_{(A,B)} := 2^8 \cdot \frac{(A^2+B^2-AB)^3}{A^2 B^2 C^2} \quad \text{with } C=A-B.$$

Its minimal discriminant is $2^{\alpha}A^2B^2C^2$ with $\alpha \geq -8$, and its conductor is equal to $N_{(A,B)} = 2^{\beta} \prod_{2\neq l|ABC} l$ with β bounded and $\beta=0$ if $2^4||A, \beta=1$ if $2^5|A$.

We want to translate divisorial properties of A,B,C into properties of $E_{(A,B)}$, and we begin by comparing $j_{(A,B)}$ with $N_{(A,B)}$. As usually define for $x\in\mathbf{Q}^*$: $h(x) := \mathrm{Max}\,\{0, \log|x|\} - \sum_{v_l(x)<0} v_l(x) \log l.$

We state the following :

<u>Conjecture</u> $(\underline{H})_d$: Let d be a positive number. For all $\epsilon>0$ and for all finite sets S of primes there is a constant $c(\epsilon,S)$ such that for all elliptic curves $E/\mathbf{Q}$ which are stable outside S one has :

$$h(j_E) \leq c(\epsilon,S) + (6d+\epsilon) \sum_{l|N_E} \log l.$$

The best value of d could be d=1, and we denote by $(\underline{H})$ the conjecture $(\underline{H})_1$.

Since the geometric height h(E) is (up to constants and a term bounded by $c.\log\log|j_E|$ for large $|j_E|$) equal to $1/12\, h(j_E)$ (cf. [De], p. 4) one can state instead of $(\underline{H})$.

<u>Conjecture</u> $(\underline{H})^*_d$: $h(E) \leq c(\epsilon,S) + (1/2d+\epsilon)(\sum_{l|N_E} \log l).$

An easy consequence of $(\underline{H})_d$ is Szpiro's conjecture

$$(\underline{SZ})_d : |\Delta_E| \leq c(\epsilon) + (6d+\epsilon).\log(N_E)$$

for all elliptic curves $E/\mathbf{Q}$ which are stable at all primes and have minimal discriminant Δ_E and conductor N_E.

<u>Remark</u> : The motivation for $(\underline{SZ})_d$ was the observation made by Szpiro that the analogous conjecture is true in the function field case with d=1 and $c= 6(2g-2)$ where g is the genus of the function field. In fact $(\underline{H})_1$ too is true in this case. It is an easy consequence of Hurwitz's genus formula, the constant c can be chosen as $c(\epsilon,S) = 6(2g-2+\sum_{p\in S} \deg(p)).$

We now list some consequences of $(\underline{H})_d$ using the special curves $E_{(A,B)}$. The most important application perhaps is the connection with :

Conjecture $(\underline{ABC})_d$: Let d be a positive number. For all $\epsilon>0$ there is a constant $c(\epsilon)$ such that for all relatively prime integers A and B and $C := A-B$ one has : $\mathrm{Max}\{|A|,|B|,|C|\} \leq c(\epsilon)(\prod_{l|ABC} l)^{(d+\epsilon)}$. Again one hopes that $d=1$ is the best choice and denotes by $(\underline{ABC})$ the conjecture $(\underline{ABC})_1$. In this precise formulation the conjecture goes back to Masser and Oesterlé. It is true in the function field case, for a higher dimensional analogue cf. [BR-Ma]. One should mention that $(\underline{ABC})$ follows from a conjecture of Vojta of "Nevanlinna-type" (cf. [Vo]).

Proposition 1. $(\underline{H})_d$ implies $(\underline{ABC})_d$

Proof : Take A and B relatively prime and apply $(\underline{H})_d$ to $E_{(A,B)}$:

$$h(2^8.\frac{(A^2+B^2-AB)^3}{A^2B^2C^2}) \leq c(\epsilon,2) + (6d+\epsilon)(\sum_{l|ABC} \log l).$$

If the order of magnitude of the absolute values of A, B and C is the same then $2(\log|A| + \log|B| + \log|C|) = c + 6\log|A|$. If (let's say) $|B| \ll |A|$ then $|C|$ has the same order of magnitude as $|A|$ and $|j_{(A,B)}|$ is equal to $c.|A|^2$, and so we get in both cases :

$$6.\mathrm{Max}\{\log|A|,\log|b|,\log|c|\} \leq c(\epsilon) + (6d+\epsilon)(\sum_{l|ABC} \log l).$$

Remark : It is easily seen that $(\underline{ABC})$ implies $(\underline{H})$, and hence is equivalent with $(\underline{H})$.

$(\underline{ABC})$ implies a lot of other conjectures, for instance Hall's conjecture and the more general conjecture of Lang and Waldschmidt about the absolute value of integral solutions of $aX^m+bY^n=k$: One expects that $|x|^{1/n} \leq c(\epsilon)|k|^{(1+\epsilon)}$.

We are interested in solutions of Fermat-type equations, and instead of applying $(\underline{ABC})$ we prefer to use $(\underline{H})_d$ directly (in fact $(\underline{SZ})_d$ would be strong enough).

Proposition 2. Let a_1,a_2,a_3 be fixed relatively prime integers and assume that $(\underline{H})_d$ holds. If the equation $a_1Z_1^n-a_2Z_2^n= a_3Z_3^n$ has nontrivial solutions for infinitely many natural numbers n then $a_1\pm a_2=\pm a_3$; in other words : $(\underline{H})_d$ implies the asymptotic Fermat conjecture.

Proof : Given a solution (z_1,z_2,z_3) of $a_1Z_1^n-a_2Z_2^n= a_3Z_3^n$ we look at $E_{(a_1z_1^n,a_2z_2^n)}$ and, assuming $(\underline{H})_d$ we get :

$$|a_1a_2a_3|^2 \cdot |z_1z_2z_3|^{2n} \le c \cdot |a_1a_2a_2z_1z_2z_3|^{6d+1}$$

and hence $|z_i|=1$ for n sufficiently large.

Properties of torsion points of $E_{(A,B)}$.

We now exploit the information about points of finite order of $E_{(A,B)}$ one gets from reduction theory of elliptic curves and especially from the theory of Tate curves.

We fix a prime $p\geq 5$ and denote by E_p the points of order dividing p of the elliptic curve E.

K_p is the field obtained by adjoining the coordinates of points in E_p to $\mathbb{Q}$, and $\varsigma_{E,p}$ is the representation of $G_{\mathbb{Q}}$ given by the action on E_p. The properties of $\varsigma_{E,p}$ we need can be found in [Se 1]; especially we have to look closer to the restriction of $\varsigma_{E,p}$ to an inertia group G_p in $G_{\mathbb{Q}}$ of the prime p : Assume that E is stable at p. If E has good supersingular reduction modulo p then K_p is tamely ramified in p. If E has ordinary reduction modulo p then $\varsigma_{E,p|G_p} = \begin{bmatrix} \chi & * \\ 0 & 1 \end{bmatrix}$, where χ is the cyclotomic character. So $K_p = K_t(\sqrt[p]{u})$ with $K_t/\mathbb{Q}$ tamely ramified in p and $u \in K_t$.

Definition : $K_p/\mathbb{Q}$ is "moderately" ramified in p if u is a $\mathfrak{p}$-adic unit for a prime $\mathfrak{p}|p$ of $\overline{\mathbb{Q}}$ corresponding to G_p, and $\varsigma_{E,p}$ is finite at p if $K_p/\mathbb{Q}$ is moderately ramified in p.

We have the

Lemma. Assume that E is stable at p. Then $\varsigma_{E,p}$ is finite at p if and only if $\mathrm{Min}\{0, v_p(j_E)\} \equiv 0 \bmod p$.

Now assume that l is a prime different from p and denote by $G_{l,i}$ the i-th ramification group of l with respect to $K_p/\mathbb{Q}$. The behaviour of $\varsigma_{E,p}$ in l is described by the number

$$n_l(\varsigma_{E,p}) := |G_{l,o}|^{-1} \cdot (\Sigma |G_{l,i}| \operatorname{codim}_{\mathbb{Z}/p}(E_p^{G_{l,i}})).$$

Define : $N_{\varsigma_{E,p}} := \prod_{p\neq l} l^{n_l(\varsigma_{E,p})}$.

We are now ready to state :

Proposition 3. For a prime $p \geq 5$ and a finite set S of primes not containing 2 and p the following assertions are equivalent :

1) There exists a nontrivial solution of $a_1 Z_1^p - a_2 Z_2^p = a_3 Z_3^p$ with $\gcd(a_1, a_2, a_3) = 1$ and $l | a_1 a_2 a_3$ only if $l \in S$.

2) There exists an elliptic curve $E/\mathbb{Q}$ satisfying :

   i) $E(\mathbb{Q})$ contains all points of order 2.
   ii) E is stable at all primes, and
   iii) the minimal discriminant of E is equal to $\Delta_E = 2^{-8} d_1 d_2^p$ with $l | d_1$ only if $l \in S$.

3) There is an elliptic curve $E/\mathbb{Q}$ satisfying :

   i) $E(\mathbb{Q})$ contains all points of order 2,
   ii) $N_{\varsigma_{E,p}} = 2 . \prod_{l \in S} l^{\delta_l}$ with $\delta_l \leq 1$, and
   iii) E is stable and $\varsigma_{E,p}$ is finite at p, and $v_2(j_E) \equiv 8 \bmod p$.

Proof : 3) is a translation of 2) into the language of representations. Assume 1) and take a relatively prime solution (z_1, z_2, z_3), assume without loss of generality that $2 | z_1$ and $a_2 z_2^p \equiv 1 \bmod 4$. Then $E_{(a_1 z_1^p, a_2 z_2^p)}$ satisfies all conditons of 2).

Assume 2). Then E has a Weierstrass equation of the form $Y^2 = X(X-A)(X-B)$, and the stability conditions imply : $2 | A$, $\gcd(A,B) = 1$ and for $l \notin S$ (hence $l \nmid d_1$) it follows that $v_l(A) \equiv v_l(B) \equiv v_l(A-B) \bmod p$, and so $(A, B, A-B)$ is a solution of an equation of the predicted type.

A variant of proposition 3 is obtained by assuming that $2^4 || a_1$ in 1), and then one has to assume in 2) and 3) that $K_p/\mathbb{Q}$ is unramified in 2.

The special case $S = \emptyset$ relates Fermat's equation $Z_1^p - Z_2^p = Z_3^p$ with elliptic curves with very special properties :

Corollary : Fermat's conjecture is false for a prime $p \geq 5$ if and only if there is an elliptic curve $E/\mathbb{Q}$ with $u_2(j_E) \equiv 8 \bmod p$ which is stable in all primes, has all points of order 2 in $E(\mathbb{Q})$ and the representation $\varsigma_{E,p}$ is finite at p and has $N_{\varsigma_{E,p}} = 2$.

§2. Modular elliptic curves.

An elliptic curve E is modular if there is a non constant morphism $\varphi : X_o(N) \longrightarrow E$, where

$$X_o(N) \otimes \mathbb{C} = \mathbb{H}^*/\Gamma_o(N) \quad \text{with}$$

$$\mathbb{H}^* = \{z \in \mathbb{C}; \operatorname{im}(z)>0\} \cup \mathbb{Q} \cup \{i.\infty\}$$

and
$$\Gamma_o(N) = \{\begin{pmatrix} a & b \\ c & d \end{pmatrix} \in Sl(2,\mathbb{Z}); \quad c \equiv 0 \bmod N\}.$$

φ is called a modular parametrization of E.

Due to Carayol we can assume that N is equal to the conductor N_E of the curve E (cf. [Ca]).

The Néron differential ω_E of $E/\mathbb{Z}$ is equal to dX/F_Y where $F(X,Y) = Y^2+a_1XY+a_3Y-X^3-a_2X^2-a_4X-a_6$ is an equation for E with $a_i \in \mathbb{Z}$ and with minimal discriminant.

Then $\omega^* := \varphi^*(\omega_E)$ is a holomorphic differential on $X_o(N)$ which is, up to a sign, uniquely determined by E.

We have : $\omega^* = f_1 dq/q$ with $f_1 = c.f$ where c is in $\mathbb{Z}^{(*)}$ and f is a cusp form of weight 2 and level N : $f(z) = q + \sum_{j=2}^{\infty} a_j q^j$ with $a_j \in \mathbb{Z}$ and $q = e^{2\pi iz}$.

Moreover f is an eigenfunction under all Hecke operators, and it is a new form (since $N=N_E$), E is determined by f up to isogeny. Conversely : To a given new form f of weight 2 and level N with Fourier coefficients in $\mathbb{Z}$ there is an elliptic curve $E_f/\mathbb{Q}$ which has a parametrization $\varphi : X_o(N) \longrightarrow E_f$ with $\varphi^*(\omega_{E_f}) = c.f.dq/q$ (for theses facts we refer to [Sh]).

The famous conjecture of Taniyama-Shimura states :

Conjecture (TS) : Every elliptic curve $E/\mathbb{Q}$ is modular.

In the following we want to show how the theory of modular elliptic curves could be used perhaps to prove the height conjecture (H) as well as to show the non existence of elliptic curves with properties described in the corollary to Proposition 3 (and so to prove Fermat's last theorem).

So *assume* that $E/\mathbb{Q}$ is a modular elliptic curve with modular parametrization $\varphi : X_o(N) \longrightarrow E$ with corresponding cusp form f. Then we have the integral formula

$$\int_{\mathbb{H}^*/\Gamma_o(N)} c^2 f^2 dV = \deg(\varphi)\mathrm{Vol}(E) = \deg(\varphi)|\omega_1|^2 \mathrm{Im}(\tau)$$

where $\{\omega_1,\omega_2\}$ is a base of the lattice in $\mathbb{C}$ corresponding to E with $|\omega_1|$ minimal and $\tau = \omega_2/\omega_1 \in \mathbb{H}$, or (cf [De]) : $h(E) \leq k + 1/2\log(\deg(\varphi))$ with a universal constant k. Hence any bound for $\deg \varphi$ will give a bound for $h(E)$ and so, with

Conjecture $(\underline{D})_d$: Let S be a finite set of primes and $\epsilon>0$. Then there is a constant $c(\epsilon,S)$ such that for all modular elliptic curves which are stable outside S there is a modular parametrization φ with $\deg(\varphi) \leq c(\epsilon,S)(\prod_{l|N_E} l)^{(d+\epsilon)}$.

We get :

Proposition 4. $((\underline{TS}) + (\underline{D})_d)$ implies $(\underline{H})_d$ and hence the asymptotic Fermat conjecture.

As far as I know the best estimate for $\deg(\varphi)$ is due to Mestre and Oesterle : $\deg(\varphi) \leq 3^{N_E}$ for $N_E \in \mathbb{R}$.

The next conjecture is a special case of a very far reaching conjecture of Serre about connections between 2-dimensional odd representations of $G(\overline{\mathbb{Q}}/\mathbb{Q})$ over finite extensions of $\mathbb{Z}/p$ and modular forms over such fields. For the exact formulation of this conjecture we refer to [Se 4].

So assume that p is a prime ≥ 5 and that ς is an irreducible representation of $G(\overline{\mathbb{Q}}/\mathbb{Q})$ with values in $Gl(2,\mathbb{Z}/p)$ such that $\det(\varsigma)$ is equal to the cyclotomic character χ, and that ς is finite at p. To have a short expression we say : ς is of finite *elliptic type* if ς satisfies the conditions imposed above.

Recall : If $E/\mathbb{Q}$ has no $\mathbb{Q}$-rational isogeny of degree p, is stable in p and has $\mathrm{Min}\{0,v_p(j_E)\} \equiv 0 \bmod p$ then $\varsigma_{E,p}$ is of finite elliptic type.

Now we can state

Conjecture S : Assume that ς is of finite elliptic type with prime-to-p-conductor N_ς. Then there is a new form $f = q + \sum_{i=2}^{\infty} a_i q^i$

of weight 2 and level N_ς with a_i algebraic integers such that there is a divisor $\wp$ of p with :

For all primes $l \nmid p.N$ one has : $a_l \equiv Tr_l(\varsigma) \bmod \wp$

where $Tr_l(\varsigma)$ denotes the trace of the image under ς of a Frobenius element at l. Or shortly : ς is modular of weight 2 and level N.

We apply (S) to Proposition 3 and get :

Proposition 5. Assume that (S) holds for squarefree levels, take a prime $p \geq 5$ and a_1, a_2, a_3 as relatively prime integers with $p \nmid a_1a_2a_3$ and $2^4 || a_1$ or $2 \nmid a_1a_2a_3$.

If (z_1, z_2, z_3) is a non trivial solution of $a_1Z_1^p - a_2Z_2^p = a_3Z_3^p$ then $\varsigma_{E_{(a_1z_1^p, a_2z_2^p), p}}$ is modular of level dividing $2. \prod_{l | a_1a_2a_3} l$.

Assume that $a_1 = a_2 = a_3 = 1$. Since $X_o(2)$ has genus 0 we get the :

Corollary : ((S)) for stable elliptic curves implies Fermat's conjecture.

A trick of B. Mazur allows to use Proposition 5 to prove results about the asymptotic Fermat conjecture too :

Proposition 6 (Mazur). Assume that (S) holds for stable elliptic curves. Let a_1, a_2, a_3 be relatively prime integers with $2^4 || a_1$ or $2 \nmid a_1a_2a_3$. If the equation $a_1Z_1^p - a_2Z_2^p = a_3Z_3^p$ has non trivial solutions for infinitely many primes p then there are a_1', a_2' and a_3' in $\mathbb{Z}$ with $a_1' - a_2' = a_3'$ and $l | a_1'a_2'a_3'$ only if $l | a_1a_2a_3.2$.

Example : Assume that (S) holds for stable elliptic curves. The the equation $2^4Z_1^p - Z_2^p = q.Z_3^p$ with a prime q has solutions for infinitely many primes p if and only if $q = 17$.

Ribet's result.

The conjecture (S) is so strong that it implies (TS). For a proof cf. [Se 4]. Though it is supported by interesting numerical examples due to Mestre and Serre it seems to be out of reach nowadays. So perhaps it is more promising to assume (TS) and then to prove (S) for representations coming from modular elliptic curves using all the deep information about $X_o(N)$ one has in the case that interests if one wants to prove something about $E_{(A,B)}$, namely : N is squarefree.

In this case $J_o(N)/\mathbb{Z}$ is fairly well known thanks to Ribet ([Ri 1]) who describes $\mathrm{End}(J_o(N)) \otimes \mathbb{Q}$ and Deligne, Rapoport and Mazur who determine the fibres of $J_o(N)$ in primes $l|N$. These informations give rather immediately the following results (cf. [Fr 2]) :

Assume that $E/\mathbb{Q}$ is stable and modular with even conductor N_E, minimal discriminant Δ_E and corresponding cusp form f. Assume that l and p are odd primes with $p \geq 5$ such that E has no $\mathbb{Q}$-rational isogeny of degree p.

If $v_l(\Delta_E) \equiv 0 \bmod p^k$ then there is a cusp form $g \neq f$ with $f \equiv g \bmod p^k$. Hence p^k is a congruence number for f (cf. [Ri 3]). If moreover ℓ is not congruent to one mod p (e.g. l=p) one finds such a g with level dividing N_E/l.

In order to come to the much deeper results of Ribet we translate the conditions imposed on ς in the conjecture ($\underline{S}$) into conditions for group schemes of type (p,p) as follows :

Let A be a quasi finite flat group scheme over $\mathbb{Z}$ of rank p^2 with $p.A=0$.

A is of <u>finite elliptic type</u> if the representation ς_A of $G_{\mathbb{Q}}$ induced by the action on $A(\overline{\mathbb{Q}})$ is of finite elliptic type (especially A is finite at p).

A is <u>semistable</u> in a prime l if the maximal finite flat subgroup $A_l/\mathbb{Z}_l$ of $A/\mathbb{Z}_l$ has $\mathbb{Z}/p$-rank ≥ 1 and A/A_l extends to a finite group scheme over $\mathbb{Z}_l$.

An example of a group scheme which is of finite elliptic type and semistable at all primes l of $\mathbb{Z}$ is E_p, the group of points of order p of a stable elliptic curve $E/\mathbb{Q}$ with $\mathrm{Min}\{0, v_p(j_E)\} \equiv 0 \bmod p$.

<u>Conjecture</u> ($\tilde{\underline{S}}$) : Assume that a group scheme $A \subset J_o(N)$ is of type (p,p) with $p \geq 5$. If A is semistable at all primes, of finite elliptic type and finite outside of $N'|N$ then ς_A is modular of weight 2 and level N'.

Ribet succeeded to prove (cf. [Ri 2]) :

<u>Theorem</u>. ($\tilde{\underline{S}}$) is true for prime numbers N.

It seems that he can prove ($\tilde{\underline{S}}$) in the general squarefree case too, and since the elliptic curves belonging to solutions of Fermat's equation have points of order p satisfying the conditions of $(\tilde{\underline{S}})$ with $N'=2$ we would get the remarkable :

<u>Consequence</u> : The conjecture of Taniyama-Shimura implies Fermat's conjecture, or more precisely : Assume that for $p\geq 5$ (z_1, z_2, z_3) is a nontrivial integral solution of $z_1^p - z_2^p = z_3^p$. Then $E : Y^2 = X(X-z_1^p)(X-z_2^p)$ is not a modular elliptic curve.

(*) p. 45 : I want to thank J. Oesterlé for explaining to me that this fact follows from the results of Katz-Mazur about models of $Y_0(N)/\mathbb{Z}$

BIBLIOGRAPHY

[Br-Ma] W.D. Brownawell and D.W. Masser.- Vanishing sums in function fields; Math. Proc. Cambr. (1986).

[Ca] H. Carayol.- Sur les représentations l-adiques associées aux formes modulaires de Hilbert; Ann. Sci. ENS 19 (1986), 409-468.

[De] P. Deligne.- Preuve des conjectures de Tate et de Shafarevitch (d'après G. Faltings); Sém. Bourbaki 616 (1983).

[De-Ra] P. Deligne and M. Rapoport.- Les schémas de modules de courbes elliptiques; in Modular Functions of One Variable II, L.N.M. 349.

[Fa] G. Faltings.- Endlichkeitssätze für abelsche Varietäten über Zahlkörpern; Invent. Math. 73 (1983), 349-366.

[Fr 1] G. Frey.- Rationale Punkte auf Fermatkurven und getwisteten Modulkurven; J. Reine u. Angew. Math. 331 (1982), 185-191.

[FR 2] G. Frey.- Links between stable elliptic curves and certain diophantine equations; Ann. Univ. Sarav. 1 (1986), 1-40.

[He] Y. Hellegouarch.- Courbes elliptiques et équations de Fermat; Thèse, Besançon 1972.

[Ku-La] D.S. Kubert and S. Lang.- Modular units; Grundlehren der mathematischen Wissenschaften 244 (1981).

[Ma 1] B. Mazur.- Modular curves and the Eisenstein ideal; Publ. Math. IHES (1977).

[Ma 2] B. Mazur.- Rational isogenies of prime degree; Invent. Math. 44, 1978.

[Ri 1] K. Ribet.- Endomorphisms of semi-stable abelian varieties over number fields; Ann. Math. 101 (1975), 555-563.

[Ri 2] K. Ribet.- Lecture at the 26. Arbeitstagung, Max-Planck-Institut für Mathematik, Bonn (1986).

[Ri 3] K. Ribet.- Mod p Hecke operators and congruences between modular forms; Invent. Math. 71 (1983), 193-205.

[Se 1] J.-P. Serre.- Propriétés galoisiennes des points d'ordre fini des courbes elliptiques; Invent. Math. 15 (1972), 259-331.

[Se 2] J.-P. Serre.- Letter to Mestre (1985).

[Se 3] J.-P. Serre.- Lecture at the conference : Théorie des nombres, CIRM Luminy (1986).

[Se 4] J.-P. Serre.- Sur les représentations modulaires de degré 2 de $\mathrm{Gal}(\overline{\mathbb{Q}}/\mathbb{Q})$; to appear in Duke Math. J.

[Sh] G. Shimura.- Introduction to the arithmetic theory of automorphic functions; Princeton Univ. Press (1971).

[Si] J.H. Silverman.- The arithmetic of elliptic curves; New York (1986).

[Vo] P. Vojta.- Diophantine approximation and value distribution theory; to appear.

[We] A. Weil.- Über die Bestimmung Dirichletscher Reihen durch Funktionalgleichungen; Math. Ann. 168 (1976), 149-156.

Gerhard FREY
Fachbereich Mathematik
der Universität des Saarlandes
D-6600 Saarbrücken
R.F.A.

POINCARE SERIES, KLOOSTERMAN SUMS, TRACE FORMULAS, AND AUTOMORPHIC FORMS FOR GL(n)

Solomon FRIEDBERG

ABSTRACT

Poincaré series on GL(n) were first introduced by Bump, the author, and Goldfeld, who studied them in the case n=3 ("Poincaré series and Kloosterman sums for $SL(3,\mathbb{Z})$", Acta Arithmetica, to appear). In this paper they are studied for arbitrary n.

First, in an expository section, the GL(2) case is reviewed in detail. In particular, Poincaré series for GL(2) are defined, their Fourier and spectral expansions are given, the Kloosterman zeta function is introduced, and Selberg's proof that $\lambda_1 \geq 3/16$ for congruence subgroups - which uses Weil's algebreo-geometric bound on the classical (GL(2)) Kloosterman sum - is sketched.

Then, the GL(n) case is discussed : Poincaré series on GL(n) are defined, their Fourier and spectral expansions are presented, and these are used to give a GL(n) version of the Bruggeman-Kuznietsov trace formula, relating the Fourier coefficients of the GL(n) cusp forms to certain exponential sums - the "GL(n) Kloosterman sums". Properties of these sums, including an algebreo-geometric bound based on Deligne's work on trigonometric sums, are also mentioned. Some of these results - the Fourier expansion of the GL(n) Poincaré series in terms of the GL(n) Kloosterman sums, and the algebreo-geometric bound - are taken from the author's forthcoming paper "Poincaré series for GL(n) : Fourier expansion, Kloosterman sums, and algebreo-geometric estimates".
AMS Classification : Primary 11 F 55, 11 F 72, 11 L 05 Secondary 11 L 40.

1 - Introduction.

Hecke eigenvalues of nonholomorphic automorphic forms on GL(2), at both the finite and infinite primes, are closely related to certain exponential sums, called (GL(2)) Kloosterman sums. This connection is visible, for example, in the trace formula of Bruggeman [B] and Kuznietsov [K], which says roughly that

$$\sum_c \frac{S(m,n;c)}{c}\, h\left(\frac{\sqrt{mn}}{c}\right) = \sum_j \overline{a_j(m)}a_j(n)\, \tilde{h}\,(\lambda_j) \tag{1.1}$$

$$+ \text{ finite sum of directly manageable terms}$$

where

$$S(m,n;c) = \sum_{\substack{a=1\\(a,c)=1}}^{c} \exp(2\pi i(ma+n\bar{a})/c) \qquad a\bar{a}\equiv 1 \bmod c \tag{1.2}$$

is the GL(2) Kloosterman sum, $a_j(m)$ is the mth Fourier coefficient of a GL(2) cusp form with Laplacian eigenvalue λ_j, h is a test function (with compact support, say), and $\tilde{h}$ is a certain integral transform of h. This may be proven by using the theory of Poincaré series on GL(2), a theory initiated by Petersson [P], and greatly developed by Selberg [Se 1].

In fact, GL(2) Poincaré series serve as valuable tools in a wide variety of contexts, allowing one to extract information on both automorphic forms and Kloosterman sums. For example, Poincaré series play a key role in Hafner's result that a positive proportion of the zeroes of the L-series attached to a cusp form lie on the critical line [H1] [H2] [H3], as well as in the large sieve inequalities for the Fourier coefficients $a_j(m)$ of Deshouillers and Iwaniec [DI]. In the other direction, Kuznietsov [K] has used (1.1) to give good bounds on sums of Kloosterman sums

$$\sum_{c<X} S(m,n;c)/c = O_{\epsilon,m,n}(X^{1/6+\epsilon}),$$

a result later obtained directly from the Poincaré series by Goldfeld and Sarnak [GS]. GL(2) Poincaré series also intercede in other analytic problems, cf. Goldfeld [G], Sarnak [Sa].

In this paper we shall present a theory of Poincaré series on GL(n), relating automorphic forms on GL(n) to certain "GL(n) Kloosterman sums". First, in section 2, which is expository, we review in some detail the GL(2) case; this will help to put the general case in perspective. In particular, Poincaré series for GL(2) are defined, their Fourier and spectral expansions are given, the Kloosterman zeta function is introduced, and Selberg's proof that $\lambda_j \geq 3/16$ for congruence subgroups - which uses Weil's algebreo-geometric bound on $S(m,n;c)$ - is sketched. These results have been known to the experts for some time, but have only recently appeared in print. Then, in section 3, we discuss Poincaré series on GL(n), explain how the "GL(n) Kloosterman sums" arise in their Fourier expansions, and how they are related to the GL(n) automorphic forms. In particular, we give a GL(n) analogue of (1.1). We also give an algebreo-geometric estimate for certain of the GL(n) Kloosterman sums; it implies that the corresponding partial Kloosterman zeta function has region of absolute convergence corresponding to the generalized Ramanujan conjecture when $n=3$, but surpassing it when $n>3$. Ultimately, these GL(n) Poincaré series should allow one to extend most of the GL(2) theorems mentioned above to higher rank cases.

The Poincaré series for GL(n) were first defined by Bump, the author, and Goldfeld in [BFG], and they were studied there in full detail when $n=3$. For general n, we are summarizing results from [Fl], plus some additional observations. I would like to thank Bump and Goldfeld for helpful conversations, and also the Institute des Hautes Etudes Scientifiques, for providing a congenial working environment during the 1985-1986 academic year. Research supported by a NATO Postdoctoral Fellowship in Science, and by NSF grant DMS-8503319.

2 - GL(2) Poincaré series.

For simplicity, we shall stick to the case of the full modular group $\Gamma = SL(2,\mathbb{Z})$; the situation for congruence subgroups is discussed in detail in [DI]. Recall that the Hilbert space $L^2(\Gamma\backslash\mathfrak{h})$ consists of square integral Γ-invariant functions on the upper half plane $\mathfrak{h}$, with inner product

$$\langle f,g\rangle = \int_{\Gamma\backslash\mathfrak{h}} f(z)\overline{g(z)}\, y^{-2}dxdy.$$

A function f in $L^2(\Gamma\backslash\mathfrak{h})$ has both a Fourier expansion

$$f(z) = \Sigma\, a(n,y)e(nx) \qquad e(x) = \exp(2\pi ix) \tag{2.1}$$

as $f(z+1) = f(z)$, and a spectral expansion, based on the decomposition

$$L^2(\Gamma\backslash\mathfrak{h}) = L^2_{cusp}(\Gamma\backslash\mathfrak{h}) \oplus L^2_{Eis}(\Gamma\backslash\mathfrak{h}) \oplus L^2_{res}(\Gamma\backslash\mathfrak{h}).$$

Here, L^2_{cusp} is the subspace spanned by the cuspidal automorphic forms, L^2_{Eis} is spanned by integrals of Eisenstein series, and L^2_{res} is spanned by the residues of Eisenstein series (in this case precisely $\mathbb{C}$). This will be explained further below.

The Poincaré series are functions in $L^2(\Gamma\backslash\mathfrak{h})$, specially constructed so that both these Fourier and spectral expansions are fully computable. Given m a positive integer, z in $\mathfrak{h}$, and ν in $\mathbb{C}$ with $Re(\nu)>1$, the GL(2) Poincaré series is given by

$$P_m(z;\nu) = \frac{1}{2} \sum_{\gamma\in\left(\begin{smallmatrix}1&*\\0&1\end{smallmatrix}\right)\backslash\Gamma} Im(\gamma\circ z)^{\nu}\, e(m(\gamma\circ z)). \tag{2.2}$$

This is well defined since $e(1)=1$. The same construction with $m=0$ would give the Eisenstein series $E(z,\nu)$; however, in contrast to the present case, $E(z,\nu)$ is <u>not</u> L^2 (the GL(n) Poincaré series of section 3, specialized to $n=2$, are slightly more general than the one given here. One can also construct Poincaré series involving various Whittaker functions, as in [Ne], [Ni]; the essential features of the discussion below do not change).

We turn to the Fourier and spectral expansions of P. The Fourier expansion (2.1) of the Poincaré series is expressed by ([Ni], [DI]).

<u>Theorem 2.1</u>. Let $\mathrm{Re}(\nu)>1$. Then

$$\int_0^1 P_m(z;\nu)e(-nx)dx = \delta_{mn}\, y^{\nu} e(niy) + \sum_{c=1}^{\infty} \frac{S(m,n;c)J(y,\nu,m,n,c)}{c^{2\nu}}$$

where $S(m,n;c)$ is the Kloosterman sum (1.2), and J is the integral

$$J(y,\nu,m,n,c) = y^{\nu} \int_{-\infty}^{\infty} (x^2+y^2)^{-\nu} e(-nx - \frac{m}{c^2(x+iy)})dx.$$

In order to give the spectral expansion of P, let f_j $(j=1,2,\ldots)$ be an orthonormal basis of cusp forms, so each f_j is by definition an eigenfunction of the Laplacian :

$$\Delta f_j = \lambda_j f_j \qquad \Delta = -y^2(\frac{\partial^2}{\partial x^2} + \frac{\partial^2}{\partial y^2}).$$

This differential equation implies that the Fourier coefficients $a_j(n,y)$ [cf. (2.1)] of the f_j are given by

$$a_j(n,y) = a_j(n)\, y^{1/2} K_{ir_j}(2\pi|n|y) \qquad (n\neq 0) \tag{2.3}$$

where $\lambda_j = \frac{1}{4} + r_j^2$, and K is a modified Bessel function. Since $\Gamma = SL(2,\mathbb{Z})$, it is known that the r_j are real, i.e. $\lambda_j \geq \frac{1}{4}$, but for congruence subgroups this is an open conjecture (cf. [Se 1], [V]). We shall explain why $\lambda_j \geq \frac{3}{16}$ below; the same proof works for congruence subgroups as well.

It is easy to see that the projection of P onto L^2_{res} is zero. Hence, its spectral expansion is ([Kub])

$$P_m(z;\nu) = \sum_j \langle P,f_j\rangle f_j + \frac{1}{4\pi}\int_{-\infty}^{\infty} \langle P,E(.,\tfrac{1}{2}+ir)\rangle\, E(z,\tfrac{1}{2}+ir)dr, \tag{2.4}$$

where

<u>Theorem 2.2</u>. Let $\mathrm{Re}(\nu)>1$. Then

(1) $\langle P_m(.;\nu),f_j\rangle = \overline{a_j(m)}\,\pi^{1/2}(4\pi m)^{1/2-\nu}\,\Gamma(\nu-\frac{1}{2}+ir_j)\Gamma(\nu-\frac{1}{2}-ir_j)/\Gamma(\nu)$

(2) $\langle P_m(.;\nu),E(.,\frac{1}{2}+ir)\rangle = 2\pi^{1+ir}m^{ir}\sigma_{-2ir}(m)(4\pi m)^{1/2-\nu}\frac{\Gamma(\nu-\frac{1}{2}+ir)\Gamma(\nu-\frac{1}{2}-ir)}{\Gamma(\nu)\Gamma(\frac{1}{2}+ir)}.$

Theorem 2.2 follows by "unfolding" the integral over $\Gamma\backslash\eta$ into one over a strip; for (2), it is also necessary to use the Fourier expansion of $E(z,\nu)$, in terms of the divisor function σ. Combining this with (2.4), one can show that

<u>Corollary 2.3</u>. $P_m(z;\nu)$ has a meromorphic continuation to all complex ν, with simple poles at

$$\nu = \frac{1}{2} \pm ir_j$$

corresponding to the cusp forms f_j (provided $\sum_{k \text{ s.t.} \lambda_k=\lambda_j} a_k(m)\neq 0$).

The Bruggeman-Kuznietsov trace formula (1.1) can now be derived as follows. As Poincaré series are L^2, it makes sense to consider the inner product

$$\langle P_m(.;u),P_n(.;\bar{\nu})\rangle = \int_0^\infty \int_0^1 P_m(z;u)y^{\nu-2}e(-n\bar{z})dxdy.$$

This integral may be computed in two ways. On the one hand, we may use Theorem 2.1 to compute the x integral, obtaining Kloosterman sums and an integral in y. On the other, we may substitute the spectral expansion of P, and use Theorem 2.2 (twice), to obtain Fourier coefficients and gamma functions. The resulting equality is useful in its own right; it implies for example that the <u>Kloosterman zeta function</u>

(2.5) $$Z_{m,n}(\nu) = \sum_{c=1}^{\infty} \frac{S(m,n;c)}{c^{2\nu}} \qquad \mathrm{Re}(\nu) >> 0$$

has a meromorphic continuation to $\mathbb{C}$, with simple poles at the lines $\nu = \frac{1}{2} \pm ir_j$, of residue (proportional to) $\sum_{k \text{ s.t.} \lambda_k=\lambda_j} \overline{a_k(m)}a_k(n)$. The final form of (1.1) follows by setting $u=\nu=1+it$, and integrating in t against a cleverly chosen function of t. The "finite sum of directly manageable terms" arises from the continuous spectrum.

Using the Riemann hypothesis for curves, Weil [W] showed that

$$|S(m,n;c)| = O(c^{1/2+\epsilon}).$$

It follows that $Z_{m,n}(\nu)$ is holomorphe in $\mathrm{Re}(\nu) > \frac{3}{4}$. By the remark above, this implies $\lambda_j \geq \frac{3}{16}$. This is Selberg's proof [Se 1].

To close this section, we remark that there are numerous relations between (1.1) and the Selberg trace formula. Indeed, Iwaniec has observed that (1.1) can be used to derive the Selberg Trace Formula. Further, Zagier has noted that the Bruggeman-Kuznietsov formula follows directly from Selberg's identity [Se 2]

$$\sum_j h_k(r_j)f_j(z)\overline{f_j(z')} = \sum_{\gamma\in\Gamma} k(z,\gamma z') - \frac{3}{\pi} h_k(\tfrac{i}{2})$$
$$- \frac{1}{4\pi}\int_{-\infty}^{\infty} E(z,\tfrac{1}{2}+ir)E(z',\tfrac{1}{2}-ir)h_k(r)dr$$

(the starting point for the STF), by looking at integrals of the $(m,n)^{th}$ Fourier coefficient. To carry this out, the same methods are employed as those used in the proofs of Theorems 2.1 and 2.2 above. Zagier's approach may also be carried over to GL(n), by modifying the ideas used to study the GL(n) Poincaré series below.

3 - GL(n) Poincaré series.

To define the GL(n) Poincaré series, let H be the homogeneous space

$$H = GL(n,\mathbb{R})/\mathbb{R}^*O(n,\mathbb{R}),$$

equipped with its natural left $GL(n,\mathbb{R})$ action. Every coset in H has a unique representative

$$\tau = uy$$

where y is now a diagonal matrix

$$y = \mathrm{diag}(y_1\cdots y_{n-1},\ldots,y_1y_2,y_1,1) \qquad y_i>0$$

and u is in the subgroup of upper triangular unipotents $\mathcal{U}$. To generalize the $\mathrm{Im}(z)^\nu$ of (2.2) set

$$I_\nu(\tau) = \prod_{i,j=1}^{n-1} y_i^{h_{ij}\nu_j} \qquad \nu = (\nu_1,\ldots,\nu_{n-1})$$

where the ν_j $(j=1,\ldots,n-1)$ are complex, and

$$h_{ij} = \begin{cases}(n-i)j & 1\le j\le i\\ (n-j)i & i\le j\le n-1\end{cases}.$$

As for $e(mz)$, let θ_1 be a character of $\mathcal{U}$ trivial on $\Gamma_\infty = \Gamma\cap\mathcal{U}$, where $\Gamma = SL(n,\mathbb{Z})$, and $\underline{E} : H \longrightarrow \mathbb{C}$ be a measurable function transforming as

$$\underline{E}(u\tau) = \theta_1(u)\underline{E}(\tau) \qquad u\in\mathcal{U},\ \tau\in H.$$

For convenience, we also require $\underline{E}$ to be bounded, though other possibilities may be treated with only minor changes.

Definition [BFG]. The GL(n) Poincaré series is given by

$$P_{\underline{E}}(\tau;\nu) = \sum_{\gamma\in\Gamma_\infty\backslash\Gamma} I_\nu(\gamma\circ\tau)\underline{E}(\gamma\circ\tau) \qquad \mathrm{Re}(\nu_i) > \frac{2}{n}.$$

It is also possible to include a cocycle here; cf. [F1].

The notion of the Fourier expansion of a Γ-invariant function f on H is more complicated when n>2. Suffice it to say that such an f may be expressed in terms of its Fourier coefficients

$$\int_{\Gamma_\infty\backslash\mathcal{U}} f(\tau)\, \overline{\theta_2(u)}\, du \qquad [\tau=uy]$$

where $\theta_2 : \Gamma_\infty\backslash\mathcal{U} \longrightarrow \mathbb{C}$ is again a character [PS] [Sh] [F2]. In order to give the evaluation of this integral when $f(\tau) \doteq P_{\underline{E}}(\tau;\nu)$, we first introduce GL(n) analogues of the Kloosterman sum (1.2).

Let W be the Weyl group of GL(n), thought of as the subgroup of permutation matrices of GL(n,$\mathbb{Z}$). Recall that there is a decomposition of GL(n,$\mathbb{R}$) into Bruhat cells G_w

$$GL(n,\mathbb{R}) = \bigcup_{w\in W} G_w \qquad \text{(disjoint union)}$$

where

$$G_w = \mathcal{U}Dw\mathcal{U} \qquad \text{(D= diagonal matrices)},$$

and that in the Bruhat decompositions

$$\gamma = b_1\, c\, w\, b_2 \qquad b_1,\ b_2\in\mathcal{U},\ c\in D$$

of γ in G_w, c (though not in general b_1,b_2) is uniquely determined by γ. Set also

$$\mathcal{U}_w^+ = w^{-1}\mathcal{U}\, w \cap \mathcal{U}, \quad \mathcal{U}_w^- = w^{-1}\, {}^t\mathcal{U}\, w \cap \mathcal{U}.$$

Definition [F1]. Given w in W, $\theta_1,\theta_2 : \Gamma_\infty\backslash\mathcal{U} \longrightarrow \mathbb{C}$ as above, and c a diagonal matrix of the form

(3.1) $\qquad c = \mathrm{diag}(\mathrm{sgn}(w)/c_{n-1}, c_{n-1}/c_{n-2}, \ldots, c_2/c_1, c_1) \qquad c_i\in\mathbb{Z},$

the GL(n) Kloosterman sum is given by

$$S(\theta_1,\theta_2;c,w) = \sum_{\substack{\gamma\in\Gamma_\infty\backslash\Gamma\cap G_w/\mathcal{U}_w^-\cap\Gamma \\ \gamma=b_1cwb_2}} \theta_1(b_1)\theta_2(b_2)$$

when

$$(3.2) \qquad \theta_1(cwuw^{-1}c^{-1}) = \theta_2(u) \qquad u\in\mathcal{U}_w^+$$

and zero otherwise.

The sum here is well defined (independent of the choice of Bruhat decomposition $\gamma = b_1c\ wb_2$) exactly when (3.2) holds. Further, only c of type (3.1) arise in the decompositions of γ in Γ. When $n=2$, $w=\begin{pmatrix} & 1 \\ 1 & \end{pmatrix}$, this reduces to the classical Kloosterman sum.

Let now V be the finite group

$$V = \Gamma \cap D$$

and

$$\theta_2^v(u) = \theta_2(vuv) \qquad v\in V, \quad u\in\mathcal{U}.$$

Then the Fourier coefficient of $P_{\underline{E}}$ is given by ([Fl], Theorem A)

<u>Theorem 3.1</u>. Let $\mathrm{Re}(\nu_i) > \frac{2}{n}$. Then

$$\int_{\Gamma_\infty\backslash\mathcal{U}} P_E(\tau,\nu)\overline{\theta_2(u)}du = \sum_{w\in W}\sum_{v\in V}\sum_{c_1,\ldots,c_{n-1}=1}^{\infty} \frac{S(\theta_1,\theta_2^v;c,w)J(y,\nu,\underline{E},\theta_2^v,c,w)}{c_1^{n\nu_1}\cdots c_{n-1}^{n\nu_{n-1}}}$$

where the S are the Kloosterman sums above, and

$$J(y,\nu,\underline{E},\theta_2^v,c,w) = \int_{\mathcal{U}_w^-} I_\nu(w^\circ uy)\underline{E}(cw^\circ uy)\overline{\theta_2^v(u)}\,du.$$

(Haar measures are chosen so that $\Gamma_\infty\backslash\mathcal{U}$ and $\Gamma_\infty\cap\mathcal{U}_w^-\backslash\mathcal{U}_w^-$ have measure 1).

When $n=3$, this is (in more abstract notation) Theorem 5.1 of [BFG]. Simultaneously, adelic versions of Theorem 3.1 have been given by I. Piatetski-Shapiro (personal communication) and G. Stevens [St]. The key idea in all proofs is the Bruhat decomposition.

As for the spectral expansion of P, write

$$\theta_1(u) = e(\Sigma\,\alpha_i u_{i,i+1}) \qquad \alpha_i\in\mathbb{Z}.$$

If $\alpha_1\ldots\alpha_{n-1}\neq 0$, then $P_{\underline{E}}(\Gamma;\nu)$ is in $L^2(\Gamma\backslash H)$. Its spectral expansion is then based on the decomposition

$$L^2(\Gamma\backslash H)= L^2_{cusp}(\Gamma\backslash H) \oplus L^2_{Eis}(\Gamma\backslash H) \oplus L^2_{res}(\Gamma\backslash H).$$

One can easily show that the projection of P onto L^2_{res} is zero, due to the lack of θ_1-Whittaker models. Hence in order to make this decomposition explicit, we must compute the invariant inner product of P against either a GL(n) cusp form or a GL(n) Eisenstein series (cf. (2.4)). Now according to Shalika's multiplicity one theorem [Sh], for any GL(n) automorphic form Φ (cuspidal or not) and <u>nondegenerate</u> character θ_1 (i.e. $\alpha_1\ldots\alpha_{n-1}\neq 0$),

$$\int_{\Gamma_\infty\backslash\mathcal{U}} \Phi(uy)\,\overline{\theta_1(u)}\,du= a_\Phi(\theta_1)W_\lambda(|\alpha|y) \tag{3.3}$$

where $a_\Phi(\theta_1)$ in $\mathbb{C}$ is the "θ_1-Fourier coefficient" of Φ, $|\alpha|y$ is the diagonal matrix with y_i replaced by $|\alpha_{n-i}|y_i$, and W_λ is a GL(n) Whittaker function [J] depending only on the eigenvalues at ∞ of Φ (compare (2.3)). More precisely, if $\lambda=(\lambda_1,\ldots,\lambda_{n-1})$ is chosen so that Φ and I_λ have the same eigenvalues for the $GL(n,\mathbb{R})$-invariant differential operators on H, then (up to analytic continuation)

$$W_\lambda(y)= \int_{\mathcal{U}} I_\lambda(w_o\circ uy)e(-\Sigma\, u_{i,i+1})du$$

where $w_o=\text{antidiag}(1,\ldots,1)$ is the long element of W. WIth this notation

<u>Theorem 3.2</u>. Let Φ be an automorphic form for GL(n), and $Re(\nu)>1$. Then the projection of P onto the space spanned by Φ is given by

$$\langle P_{\underline{E}}(.,\nu),\Phi\rangle = \overline{a_\Phi(\theta_1)}\;\Gamma(\nu,\lambda;\theta_1,\underline{E})$$

where

$$\Gamma(\nu,\lambda;\theta_1,\underline{E})= \int_{\mathbb{R}_+^{n-1}} I_\nu(y)\underline{E}(y)\overline{W}_\lambda(|\alpha|y)d^*y$$

$$d^*y = (y_1^n\, y_2^{2n-3} \ldots y_i^{i(n-i)+1} \ldots y_{n-1}^n)^{-1}\, dy_1 \ldots dy_{n-1}.$$

Theorem 3.2 is less precise than Theorem 2.2 in that the integral $\Gamma(\nu,\lambda;\theta_1,\underline{E})$ has not been explicitly evaluated. If one chooses, say,

$$\underline{E}(\tau)=\theta_1(u)e(i|\alpha_1|y_{n-1}+\ldots+i|\alpha_{n-1}|y_1)$$

and uses a Taylor expansion, this evaluation reduces to computing the (multiple) Mellin transform of W_λ. As explained in the appendix to [F3] (cf. also [BF]), at the moment one only has an idea how to do this integral when the ν_i are specialized to lie on a specific hypersurface of dimension 2. Ultimately, the spectral expansion and Theorem 3.2 should give the meromorphic continuation of P to all ν in $\mathbb{C}^{n-1}$, and show that P has polar divisor corresponding to the cuspidal automorphic forms. At least we will know this on a hypersurface of dimension 2 (once some details are completed).

Now choose two characters θ_1, θ_2 on $\Gamma_\infty\backslash\mathcal{U}$ with $\theta_2(u)=e(\Sigma\,\beta_i u_{i,i+1})$ also nondegenerate, corresponding $\underline{E}_1, \underline{E}_2$, and u,ν in $\mathbb{C}^{n-1}$. Then proceeding as in section 2, we obtain the following GL(n) version of the Bruggeman-Kuznietsov formula :

<u>Theorem 3.3</u>. Let $\mathrm{Re}(u_i)$, $\mathrm{Re}(\nu_i) > \frac{2}{n}$. Then

$$\sum_{w\in W}\sum_{v\in V}\sum_{c_1,\ldots,c_{n-1}=1}^{\infty}\frac{S(\theta_1,\overset{v}{\theta}_2;c,w)}{c_1^{nu_1}\cdots c_{n-1}^{nu_{n-1}}}\cdot\int_{\mathbb{R}_+^{n-1}}J(y,u,\underline{E}_1,\overset{v}{\theta}_2,c,w)I_\nu(y)\underline{E}_2(y)d^*y$$

$$=\sum_{\text{cusp forms }\phi_j}\overline{a_{\phi_j}(\theta_1)}a_{\phi_j}(\theta_2)\Gamma(u,\lambda_j;\theta_1,\underline{E}_1)\Gamma(\nu,\overline{\lambda_j};\theta_2,\underline{E}_2)$$

$$+\int_{\Gamma_\infty\backslash H}\mathrm{Pr}_{\mathrm{Eis}}(P_{\underline{E}_1}(\tau;u))\;I_\nu(\tau)\;\overline{\theta_2(u)}\;\underline{E}_2(y)\;du\;d^*y$$

where the first sum on the right hand side is over an orthonormal basis of cusp forms ϕ_j, with corresponding eigenvalues λ_j as in (3.3), and $\mathrm{Pr}_{\mathrm{Eis}}$ denotes the projection onto $L^2_{\mathrm{Eis}}(\Gamma\backslash H)$.

Using Theorem 3.2, one can express the second summand on the right (which is the "directly manageable term" of (1.1) when n=2) in terms of the Fourier coefficients of Eisenstein series and multiple integrals (in λ) of the functions $\Gamma(u,\lambda;\theta,\underline{E})$. Since the non-minimal-parabolic Eisenstein series incorporating lower rank cusp forms contribute to L^2_{Eis}, this formula is complicated, and we omit it here. Also, Theorem 3.3 presumably is valid in $\mathbb{C}^{n-1}\times\mathbb{C}^{n-1}$ after continuation. Then integrating against a function of (u,ν), one may derive formulas with test functions and their transforms (we may do this now, but only within the tube domain $\mathrm{Re}(u_i)$, $\mathrm{Re}(\nu_i) > \frac{2}{n}$ or on the hypersurface mentioned above).

We turn briefly to the Kloosterman sums themselves.

Theorem 3.4 (Properties of the GL(n) Kloosterman sums).

(1) (Multiplicativity) Let c_i', c_i'' $(1 \le i \le n-1)$ be nonzero integers such that $\gcd(\Pi_i c_i', \Pi_i c_i'')=1$, and c', c'' be the corresponding diagonal matrices as in (3.1). Then there exist characters θ_2', θ_2'' of $\mathcal{U}$ such that

$$S(\theta_1,\theta_2;c'c'',w) = S(\theta_1,\theta_2';c',w)\, S(\theta_1,\theta_2'';c'',w).$$

(2) (Reduction to lower rank) If $c_1=1$ then each GL(n) Kloosterman sum is a GL(n-1) sum.

(3) (Long element sum) If the c_i are pairwise relatively prime, then $S(\theta_1,\theta_2;c,w_o)$ is a product of GL(2) sums.

(4) (Algebreo-geometric estimate) Let $w_1 = \begin{pmatrix} 0 & 1 \\ Id_{n-1} & 0 \end{pmatrix}$. Then for $1 \le j \le n-1$

$$|S(\theta_1,\theta_2;c,w_1)| = O\Big(c_j^{((n-1)^2/2j)+\epsilon} \prod_{\substack{p|c_{n-1} \\ p\nmid c_j}} p^{(n-j-1)(n-j-2)/2}\Big).$$

The algebreo-geometric estimate here is deduced by using a Plücker coordinate parametrization of the quotient space $\Gamma_\infty \backslash \Gamma \cap G_w / \mathcal{U}_w^- \cap \Gamma$ (valid for any w), an explicit Bruhat decomposition of G_{w_1}, Deligne's estimate [D] for the trigonometric sum

$$\sum_{X_i \in \mathbb{F}_p,\, X_1 \cdots X_n = 1} e((X_1+\cdots+X_n)/p),$$

and ad-hoc arguments to bound the remaining sums which arise. For a more precise version, see [Fl], Theorem C. Also, for more precise statements of parts (1) and (2), see [Fl], Propositions 2.4 and 3.6.

Recall that on GL(2) Theorem 3.3 together with the evaluation of $\Gamma(\nu,\lambda;\theta,\underline{E})$ implies that the Kloosterman zeta function $Z_{m,n}(\nu)$ of (2.5) has poles corresponding to the GL(2) cusp forms. By analogy, one expects that the <u>partial Kloosterman zeta functions</u>

$$Z_{\theta_1,\theta_2,w}(\nu) = \sum_{c_1,\ldots,c_{n-1}=1}^{\infty} \frac{S(\theta_1,\theta_2;c,w)}{c_1^{n\nu_1} \cdots c_{n-1}^{n\nu_{n-1}}} \qquad \mathrm{Re}(\nu_i) \gg 0$$

should have meromorphic continuations with polar divisors corresponding to the GL(n) cusp forms, at least for certain w.

Using the algebreo-geometric estimate above in the case $n=3$ (where it is due to Larsen [L]) it was shown in [BFG] that the region of absolute convergence of $Z_{\theta_1,\theta_2,w_1}$ is exactly that of the $GL(3)$ generalized Ramanujan conjecture, in the sense that if $Z_{\theta_1,\theta_2,w_1}$ inherited the poles of P, then Theorem 3.4, (4) would imply the conjecture (compare Selberg's proof that $\lambda_j \geq 3/16$ on $GL(2)$). However, for $n>3$ this is not true, as explained (for $n=4$) in [F3]. It seems likely that there is some connection between the poles of the Dirichlet series $Z_{\theta_1,\theta_2,w}$ and the $GL(n)$ automorphic forms, but the precise relation is not yet clear.

BIBLIOGRAPHY

[B] R.W. Bruggeman.- Fourier coefficients of cusp forms, Invent. Math. 45 (1978), 1-18.

[BF] D. Bump, S. Friedberg.- The exterior square automorphic L-functions on GL(n), preprint (1986).

[BFG] D. Bump, S. Friedberg, D. Goldfeld.- Poincaré series and Kloosterman sums for SL(3,$\mathbb{Z}$), Acta Arithmetica, to appear.

[D] P. Deligne.- Application de la formule des traces aux sommes trigonométriques, in SGA $4\frac{1}{2}$, Springer Lecture Notes 569 (1977), 168-232.

[DI] J.-M. Deshouillers, H. Iwaniec.- Kloosterman sums and Fourier coefficients of cusp forms, Invent. Math. 70 (1982), 219-288.

[F1] S. Friedberg.- Poincaré series for GL(n) : Fourier expansion, Kloosterman sums, and algebreo-geometric estimates, I.H.E.S. preprint, December 1985.

[F2] S. Friedberg.- The Fourier expansion on a congruence subgroup of SL(r,$\mathbb{Z}$), in Proceedings of the Stillwater Analytic Number Theory Conference, to appear.

[F3] S. Friedberg.- Explicit determination of GL(n) Kloosterman sums, I.H.E.S. preprint, July 1986.

[G] D. Goldfeld.- Analytic and arithmetic theory of Poincaré series, Astérisque 61 (1979), 95-107.

[GS] D. Golfeld, P. Sarnak.- Sums of Kloosterman sums, Invent. Math. 71 (1983), 243-250.

[H1] J.L. Hafner.- Explicit estimates in the arithmetic theory of cusp forms and Poincaré series, Math. Ann. 264 (1983), 9-20.

[H2] J.L. Hafner.- Zeros on the critical line for Dirichlet series attached to certain cusp forms, Math. Ann. 264 (1983), 21-38.

[H3] J.L. Hafner.- On the zeroes of Maass wave form L-functions, Bull. A.M.S. 15 (1986), 61-64.

[J] H. Jacquet.- Fonctions de Whittaker associées aux groupes de Chevalley, Bull. Soc. Math. France 95 (1967), 243-309.

[Kub] T. Kubota.- Elementary theory of Eisenstein series, John Wiley and Sons, 1973.

[K] N.V. Kuznietsov.- The Petersson conjecture for cusp forms of weight zero and the Linnik conjecture. Sums of Kloosterman sums, Math. Sbornik (N.S.) 111 (153, N° 3) (1980), 334-383.

[L] M. Larsen.- Estimation of SL(3,ℤ) Kloosterman sums, appendix to [BFG], Acta Arithmetica, to appear.

[Ne] H. Neunhöffer.- Über die analytische Fortsetzung von Poincaréreihen, Sitz. Heidelberger Akad. Wiss., 2 Abhandlung (1973), 33-90.

[Ni] D. Niebur.- A class of nonanalytic automorphic functions, Nagoya Math. J. 52 (1973), 133-145.

[P] H. Petersson.- Über die Entwicklungskoeffizienten der Automorphen Formen, Acta Math. 58 (1932), 169-215.

[PS] I. Piatetski-Shapiro.- Euler subgroups, in Lie Groups and their Representations, John Wiley and Sons (1975), 597-620.

[Sa] P. Sarnak.- Additive number theory and Maass forms, in Number Theory (New York 1982), Springer Lecture Notes 1052 (1984), 286-309.

[Se1] A. Selberg.- On the estimation of Fourier coefficients of modular forms, Proc. Symp. Pure Math. 8 (1965), 1-15.

[Se2] A. Selberg.- Harmonic analysis and discontinuous groups in weakly symmetric Riemannian spaces with applications to Dirichlet series, J. Indian Math. Soc. 20 (1956), 47-87.

[Sh] J. Shalika.- The multiplicity one theorem for GL(n), Ann. Math. 100 (1974), 171-193.

[St] G. Stevens.- Poincaré series on GL(r) and Kloosterman sums, preprint (1986).

[V] M.-F. Vigneras.- Quelques remarques sur la conjecture $\lambda_1 \geq 1/4$, in Séminaire de Théorie des Nombres de Paris 1981-1982, Birkhäuser, 321-343.

[W] A. Weil.- On some exponential sums, Proc. Nat. Acad. Sci. 34 (1948), 204-207.

Solomon FRIEDBERG
Math. Dept., U.C.S.C.
Santa Cruz CA 95064
U.S.A.

SOME RECENT RESULTS ON COMPLEX POWERS AND ZETA DISTRIBUTIONS*

Jun-Ichi IGUSA

This is an expanded and updated version of our talk given on Dec. 2, 1985 in the Séminaire de Théorie des Nombres de Paris under the title, "Complex powers and their functional equations". We have tried to state definitions precisely and results concisely. We have included some remarks not stated elsewhere. As we emphasized in the talk, a problem of Weil is the underlying theme.

1. Weil's problem.

We shall denote by $\mathcal{S}(\mathfrak{X})$ the Schwartz-Bruhat space of a locally compact abelian group $\mathfrak{X}$ and by $\mathcal{S}(\mathfrak{X})'$ its topological dual; an element of $\mathcal{S}(\mathfrak{X})'$ is called a tempered distribution in $\mathfrak{X}$. Also we shall denote by $\Omega(\mathcal{G})$ the topological abelian group of all quasicharacters, i.e., 1-dimensional representations, of a locally compact group $\mathcal{G}$; if $\mathcal{G}$ is compactly generated, then $\Omega(\mathcal{G})$ forms a complex Lie group. If $\mathcal{G}$ acts on $\mathfrak{X}$ as gx, it acts on $\mathcal{S}(\mathfrak{X})$, $\mathcal{S}(\mathfrak{X})'$ as $\phi^g(x) = \phi(gx)$, $(gT)(\phi) = T(\phi^g)$, where g, x, ϕ and T are in $\mathcal{G}$, $\mathfrak{X}$, $\mathcal{S}(\mathfrak{X})$ and $\mathcal{S}(\mathfrak{X})'$, respectively. If now $\tilde{\omega}$ is an element of $\Omega(\mathcal{G})$, we shall denote by $\mathcal{E}_{\mathfrak{X}}(\tilde{\omega})$ the subspace of $\mathcal{S}(\mathfrak{X})'$ consisting of all T such that $gT = \tilde{\omega}(g)^{-1}T$ for every g in $\mathcal{G}$. The problem of Weil [24] is to determine all elements of $\mathcal{E}_{\mathfrak{X}}(\tilde{\omega})$ in the case where $\mathfrak{X}$ is a finite dimensional vector space over a local field or an adele ring.

If F is any field and X is an affine n-space Aff^n, after Sato [17], [18] we understand by a regular prehomogeneous F-structure in X a connected reductive F-subgroup G of $GL(X)=GL_n$ with a dense orbit Y in X such that $X-Y$ is an irreducible hypersurface $f(x)=0$; we may then assume that the coefficients of $f(x)$ are in F. We observe that $f(x)$ is determined up to a factor in $F^\times = F-\{0\}$ by G. The definition implies that every g in G gives a similarity of $f(x)$, i.e., $f(gx) = \upsilon(g)f(x)$ with $\upsilon(g)$ independent of x; in particular $f(x)$ is homogeneous. Since G can be enlarged by including the center $(GL_1)1_n$ of GL_n, we sometimes assume that G contains the center if it is not already the case.

If k is a number field, i.e., a finite algebraic extension of $\mathbb{Q}$, we shall denote by k_v, k_A and $k_A^\times$ a completion of k, the adele and idele groups of k, respectively. If G or rather (G,X) is a regular prehomogeneous k-structure in X, we ask Weil's problem for $\mathcal{E}_{\mathfrak{X}}(\omega\circ\nu)$ in which either $\mathfrak{X}= X_v= k_v^n$, $\mathfrak{G}=G_v$ and ω is in $\Omega(k_v^\times)$ or $\mathfrak{X}= X_A= k_A^n$, $\mathfrak{G}=G_A$ and ω is in $\Omega(k_A^\times/k^\times)$.

2. <u>Complex powers</u>.

We shall denote k_v by K. Therefore K is either $\mathbb{R}$, $\mathbb{C}$ or a p-adic field, i.e., a finite algebraic extension of $\mathbb{Q}_p$ for some prime number p. In the p-adic case we shall denote by O_K the maximal compact subring of K, by $\pi_K O_K$ its ideal of nonunits and by q_K the number of elements of $O_K/\pi_K O_K$, i.e., $q_K= |O_K/\pi_K O_K|$. In the general case we shall denote the normalized absolute value on K by $|a|_K$; if dx is a Haar measure on K, symbolically $d(ax)= |a|_K dx$ for every a in $K^\times$. The identity component of $\Omega(K^\times)$ consists of $\omega_s(a)= |a|_K^s$ for all s in $\mathbb{C}$; if ω is arbitrary in $\Omega(K^\times)$, there exists a unique $\sigma(\omega)$ in $\mathbb{R}$ satisfying $|\omega(a)|= \omega_{\sigma(\omega)}(a)$ for every a in $K^\times$. If σ is in $\mathbb{R}$, we shall denote by $\Omega_\sigma(K^\times)$ the open subset of $\Omega(K^\times)$ defined by $\sigma(\omega)>\sigma$.

We take $f(x)$ arbitrarily from $K[x_1,\dots,x_n]-K$, ω from $\Omega_0(K^\times)$ and ϕ from $\mathcal{S}(X_K)$; then

$$\omega(f)(\phi)= \int_{X_K} \omega(f(x))\phi(x)dx,$$

where dx is a Haar measure on X_K, defines an $\mathcal{S}(X_K)'$-valued holomorphic function $\omega(f)$ on $\Omega_0(K^\times)$. A fundamental theorem states that $\omega(f)$ has a meromorphic continuation to the whole $\Omega(K^\times)$ and that if K is a p-adic field, then $\omega(f)(\phi)$ is a rational function of $\omega(\pi_K)$. This theorem was proved by Atiyah [1], Bernshtein and Gel'fand [2], Bernshtein [3] and [8]. If ω' is not a pole of $\omega(f)$, then $\omega'(f)$ has X_K as its support, hence $\omega'(f)\neq 0$. Therefore $\omega(f)$ has no zeros on $\Omega(K^\times)$. The proofs of the theorem provide some general information about the poles of $\omega(f)$. However the determination of actual poles of $\omega(f)$ is a difficult problem; we just mention that there are remarkable results by Barlet, Lichtin, Loeser, Meuser and others on this problem.

The fundamental theorem remains valid even if we replace X_K by a certain type of subsets D of X_K, i.e., $\omega(f)$ by $\omega(f_D)$ defined by

$$\omega(f_D)(\phi)= \int_D \omega(f(x))\phi(x)dx$$

with ϕ still arbitrary in $\mathcal{S}(X_K)$. If $K=\mathbb{R}$, the theorem was proved in that generality in [1], [3]. If K is a p-adic field, the generalization is due to Denef [5] and in fact by a new method. We shall later take as D a G_K-orbit in Y_K in the special case where (G,X) is a regular prehomogeneous K-structure in X.

The meromorphic tempered distribution $\omega(f_D)$ is called a complex power of $f(x)$. In the special case where K is a p-adic field, $D=X_K$, $\omega=\omega_s$ and ϕ is the characteristic function of $X(O_K)=O_K^n$ the rational function $Z(s)= \omega(f)(\phi)$ of q_K^{-s} has been computed for many $f(x)$; the results are summarized in [9]; cf. also [10]. The list of $f(x)$ contains all invariants except for those of Λ^3SL_7, Λ^3SL_8 and $Spin_{14}$ among the connected irreducible simple matrix groups with one invariant; cf. [13]. If we start from a homogeneous polynomial $f(x)$ with coefficients in k, for almost all p-adic completions $k_v=K$ the degree of $Z(s)$ in q_K^{-s} is equal to $-\deg(f)$. This theorem has recently been proved under a nondegeneracy assumption on $f(x)$ by Meuser and in the general case by Denef [6]. Actually Denef has obtained an expression for $Z(s)$ valid for almost all K in terms of what might be loosely called resolution data over $\mathbb{F}_{q_K}$.

3. Functional equations.

We shall assume that (G,X) is a regular prehomogeneous K-structure in X. Then Y_K splits into a finite number of G_K-orbits $Y_1,\dots,Y_\ell$; cf. Serre [19], III-33. We put $d= \deg(f)$, $\kappa=n/d$ and introduce ℓ complex powers $Z_1,\dots,Z_\ell$ as

$$Z_i(\omega)= (\omega\omega_{-\kappa})(f_{Y_i}),\quad 1\leq i\leq \ell.$$

As we have explained in § 2, they are $\mathcal{S}(X_K)'$-valued meromorphic functions on $\Omega(K^\times)$.

In order to prove a system of functional equations for them we shall assume that there exists a K-involution of M_n under which G is stable; equivalently we shall assume that there exists $h=\pm{}^th$ in $GL_n(K)$ such that $hGh^{-1}={}^tG$. Such an involution or an element h may always exist; it at least exists if G is a K-split group and also for $K=\mathbb{R}$. We then choose a nontrivial character $\psi=\psi_K$ of K and define Fourier transformations in $\mathcal{S}(X_K)$, $\mathcal{S}(X_K)'$ as

$$\phi^*(x)= \int_{X_K} \psi_K({}^t xhy)\phi(y)dy, \quad T^*(\phi)= T(\phi^*);$$

we shall normalize the Haar measure dx to be selfdual. In the p-adic case we shall further assume that G has only a finite number of orbits in $f^{-1}(0)$. Then there exist ℓ^2 meromorphic functions $\gamma_{ij}(\omega)$ on $\Omega(K^\times)$ such that

$$Z_i(\omega)^*= \sum_{j=1}^{\ell} \gamma_{ij}(\omega)Z_j(\omega_\kappa\omega^{-1}), \qquad 1\leq i\leq \ell;$$

and in the p-adic case $\gamma_{ij}(\omega)$ are rational functions of $\omega(\pi_K)$.

At the basis of the proof lies the fact that $Z_i(\omega)^*$ and $Z_j(\omega_\kappa\omega^{-1})$ are elements of $\mathcal{E}_{\mathfrak{X}}(\tilde{\omega})$ for $\mathfrak{X}=X_K$ and $\tilde{\omega}= (\omega_\kappa\omega^{-1})\circ\nu$. Since each Y_i is a G_K-homogeneous space, by the theory of homogeneous spaces we have the above system of equations modulo elements of $\mathcal{E}_{\mathfrak{X}}(\tilde{\omega})$ supported by $f^{-1}(0)_K$. If the number of G-orbits in $f^{-1}(0)$ is finite, so is the number of G_K-orbits in $f^{-1}(0)_K$ by Serre; and then such singular distributions do not exist for a variable ω. In the case where $K=\mathbb{R}$, $\mathbb{C}$ we can eliminate them without the above finiteness assumption. We refer the details to Sato and Shintani [18], Shintani [21] and [9].

We put

$$\gamma(G,\omega,f,h,\psi)= \begin{bmatrix} \gamma_{11}(\omega) & \cdots & \gamma_{1\ell}(\omega) \\ \cdot & & \cdot \\ \cdot & & \cdot \\ \cdot & & \cdot \\ \gamma_{\ell 1}(\omega) & \cdots & \gamma_{\ell\ell}(\omega) \end{bmatrix}$$

and call it a Γ-matrix associated with (G,X). We recall that

$$\gamma(\omega,\psi)= \gamma(GL_1,\omega,x,1,\psi)$$

is known in the Tate theroy [22]. In the typical case where $K=\mathbb{R}$ the Γ-matrix has been determined up to certain Fourier polynomials in the above-quoted papers of Sato and Shintani. It is known for $K=\mathbb{C}$ and it has been computed in some special cases.

4. Normalized Γ-matrices.

If we put $f^*(x)=f(h^{-1}x)$, there exists a polynomial $b(s)=s^d+\ldots$ of degree $d=\deg(f)$ with coefficients in $\mathbb{Q}$ satisfying

$$f^*(\partial/\partial x)f(x)^{s+1}=c(f,h)b(s)f(x)^s$$

with $c(f,h)$ in $K^\times$ for every s in $\mathbb{N}$; $b(s)$ is the Bernshtein polynomial of $f(x)$ and it is called the b-function of (G,X) by Sato. If we write

$$b(s)=\prod_\lambda (s+\lambda),$$

then all λ are in $\mathbb{Q}$ and positive by Kashiwara [14].

On the other hand if we put $Y^*=hY$, then Y^* is a K-open subset of X independent of h and ${}^tG$ is transitive on Y^*. Furthermore Y^*_K splits into ${}^tG_K$-orbits $Y^*_1,\ldots,Y^*_\ell$, which coincide up to a permutation with $hY_1,\ldots,hY_\ell$. We shall denote by $P(h)$ the permutation matrix defined by $hY_i=P(h)_{ij}Y^*_j$ with $P(h)_{ij}=1$ for $1\le i\le\ell$. The point is that if we put

$$a_K(G,\omega)=(\omega_{\kappa/2}\omega^{-1})(c(f,h)).\prod_\lambda \gamma(\omega\omega_{\lambda-\kappa},\psi)^{-1}.P(h)^{-1}.\gamma(G,\omega,f,h,\psi),$$

then $a_K(G,\omega)$ becomes independent of f, h and ψ_K. It depends on the K-conjugacy class of G in GL_n and also on the ordering $Y_1,\ldots,Y_\ell$, $Y^*_1,\ldots,Y^*_\ell$. We call $a_K(G,\omega)$ the normalized Γ-matrix of (G,X). A basic problem is to find an explicit description of $a_K(G,\omega)$.

In the special case where $\ell=1$ under certain additional conditions we have $a_K(G,\omega)=1$. This means that up to an explicit normalization

$$\gamma(G,\omega,f,h,\psi)=\prod_\lambda \gamma(\omega\omega_{\lambda-\kappa},\psi),$$

in which $\gamma(\omega,\psi)$ is the Tate Γ-factor. If $K=\mathbb{C}$, there is no condition; if $K=\mathbb{R}$, the condition is that $f(x)$ can be written as

$$f(x)=\sum_{i_1<\ldots<i_d} c_{i_1\ldots i_d}\, x_{i_1}\ldots x_{i_d}.$$

If K is a p-adic field, the conditions are that G is an irreducible K-split group and the roots of $b(s)$ are all in $\mathbb{Z}$. These conditions imply not only the finiteness of G-orbits in $f^{-1}(0)$ but also the above special expression for $f(x)$.

If we drop the assumption that G splits over K, then $a_K(G,\omega)$ need not be 1. For instance if $K=\mathbb{R}$ and if $f(x)$ is a definite quadratic form in n variables, G is the identity component of the group $Sim(f)$ of its similarities and $\omega=\omega_s(sgn)^m$ with $m=0,1$, then

$$a_{\mathbb{R}}(G,\omega)=\cos(n\pi/4)+\sin(n\pi/4)\tan((s+m)\pi/2).$$

We have $b(s)=(s+1)(s+n/2)$ in this case. Therefore the roots of $b(s)$ are in $\mathbb{Z}$ if and only if n is even. In that case the discriminant $D(f)$ of $f(x)$, which is defined in general as

$$D(f)=(-1)^{n(n-1)/2}.\det(h)$$

for $f(x)=1/2.{}^t xhx$, ${}^t h=h$, has $(-1)^{n/2}$ as its sign. Therefore G becomes an inner $\mathbb{R}$-form of an $\mathbb{R}$-split group if and only if $n\equiv 0 \bmod 4$; and then $a_{\mathbb{R}}(G,\omega)=(-1)^{n/4}$. Similar results in a few other cases can be stated uniformly as the following common theorem : If G is an inner K-form of an irreducible K-split group, $\ell=1$ and all roots of $b(s)$ are in $\mathbb{Z}$, then X can be considered as a direct sum of the underlying spaces of central division K-algebras of dimensions d_i^2 such that

$$a_K(G,\omega)=(-1)^{\sum_i (d_i-1)}.$$

We refer to [10] for the details.

5. <u>Classification of</u> (G,X).

The results in § 4 which depend on the irreducibility of the matrix group G are obtained by using the classification over $\mathbb{C}$ of all such (G,X) by Sato and Kimura [17]. We shall briefly recall their theory replacing $\mathbb{C}$ by any field F of characteristic 0 : Let X, X' denote affine F-spaces and G, G' F-subgroups of $GL(X)$, $GL(X')$; then the pairs (G,X), (G',X') are called F-isomorphic if there exist F-isomorphisms $G \xrightarrow{\sim} G'$, $X \xrightarrow{\sim} X'$, denoted by $g \longmapsto g'$, $x \longmapsto x'$, satisfying $(gx)'=g'x'$ for every g in G and x in X. We shall not distinguish F-isomorphic pairs. If $X_o=Aff^{n_o}$, G_o is an F-subgroup of GL_{n_o} and $n_o=p+q$ is a partition of n_o, then $G_o\times GL_p$ (resp. $G_o\times GL_q$) acts on $M_{n_o,p}$ (resp. $M_{n_o,q}$) as $\rho_1(g_o,g_1)x_1=g_o x_1 {}^t g_1$ (resp. $\rho_2(g_o,g_2)x_2={}^t g_o^{-1} x_2 {}^t g_2$). In this situation the classical F-isomorphism of the Grassmannians of p- and q-dimensional subspaces of X_o becomes G_o-equivariant. It follows from this fact that if we put $G_1 = \rho_1/G_o \times GL_p)$, $X_1 = M_{n_o,p}$ and $G_2 = \rho_2(G_o \times GL_q)$, $X_2 = M_{n_o,q}$, then (G_1,X_1) and (G_2,X_2) share some properties, e.g., the property of G_i having a dense orbit Y_i in X_i and further the property of Y_i being the complement of an irreducible F-hypersurface $f_i(x_i)=0$. On the other hand if (G_o,X_o) and $({}^tG_o,X_o)$ are F-isomorphic, e.g., if G_o is a connected reductive F-split group, then the relation of (G_1,X_1) and (G_2,X_2) becomes symmetric. We say that (G_1,X_1), (G_2,X_2) are neighboring pairs ("castling transforms of each other" in Sato and Kimura). We shall consider the set of all pairs (G,X), in which $X= Aff^n$ for some n and G is a connected irreducible F-split subgroup of $GL(X)$ acting transitively on the complement Y of an irreducible F-hypersurface $f(x)=0$ in X. We then say that (G,X), (G',X') in the set are equivalent if they can be included in a finite sequence of successive neighboring pairs. Each equivalence class contains a unique pair for which $\dim(X)$ is the smallest; such a pair is called reduced. According to Sato and Kimura, there are 29 types of equivalence classes. Furthermore Kimura has given in [15] a list of b-functions of all reduced pairs with one exception.

Now the number ℓ of G_F-orbits in Y_F is an invariant of the class and it is finite for $F=K$ by Serre. By definition $\ell=1$ for $K=\mathbb{C}$. We have found that $\ell=1$ for $K=\mathbb{R}$ if and only if all roots of the b-function of the reduced pair in the class are in $\mathbb{Z}$. There are 8 types of such classes. Furthermore if K is a p-adic field, we have $\ell=1$ for just more class with $(\Lambda^3 GL_7, \Lambda^3 Aff^7)$ as its reduced pair. Actually the above result is valid if, instead of $K=\mathbb{R}$ or a p-adic field, there exists a nonsplit octonion F-algebra or a nonsplit quaternion, but not octonion, F-algebra, respectively; cf. [11]. The 9 types consist of 4 types, each with infinitely many classes, and 5 types, each with only one class. Among them, in fact among the 29 types, there is only one type for which the reduced pair involves an unspecified group; it is $(\rho(G_o \times GL_d), M_d)$, in which G_o is an arbitrary connected irreducible F-split subgroup of SL_d, hence $f(x)= \det(x)$, for $d=1,2,3,\ldots$

6. Relation of $Z(s)$ for neighboring pairs.

We have introduced $Z(s)$ in § 2 if K is a p-adic field. If $K= \mathbb{R}$ (resp. $\mathbb{C}$), we similarly define $Z(s)$ with $\phi(x)= \exp(-\pi\, {}^t xx)$ (resp. $\exp(-2\pi\, {}^t\bar{x}x)$). We further normalize dx so that $Z(0)=1$. If we denote this dx by dx_K, then dx_k is the n-th power of the usual Haar measure on K, i.e., the one such that the measure of the unit disc is 2, 2π or 1 for $K=\mathbb{R}$, $\mathbb{C}$ or a p-adic field.

If now (G_1,X_1), (G_2,X_2) are the neighboring pairs in § 5 with $q \geq p$, $F=K$ and with G_i having the complement of an irreducible

K-hypersurface $f_i(x_i)=0$ in X_i as its dense orbit and if we denote the $Z(s)$ for $f_i(x_i)$ by $Z_i(s)$, then

$$(*) \qquad Z_2(s)/Z_1(s) = \pi^{(p-q)ds/2} \cdot \prod_{p<i\leq q} \Gamma((ds+i)/2)/\Gamma(i/2),$$
$$(2\pi)^{(p-q)ds} \cdot \prod_{p<i\leq q} \Gamma(ds+i)/\Gamma(i) \quad \text{or}$$
$$\prod_{p<i\leq q} (1-q_K^{-i})/(1-q_K^{-(ds+i)})$$

for $K=\mathbb{R}$, $\mathbb{C}$ or a p-adic field. In the above formula d denotes the degree of a homogeneous polynomial f_o in

$$N = \binom{n_o}{p} = \binom{n_o}{q}$$

variables such that f_1 and f_2 become the compositions of f_o and the morphisms to the Grassmannians of p- and q-dimensional subspaces of X_o.

If $K=\mathbb{C}$, then (*) can be rewritten as

$$Z_2(s)/Z_1(s) = (2\pi/d)^{(p-q)ds} \cdot \prod_{p<i\leq q} \prod_{0\leq j<d} \Gamma(s+d^{-1}(i+j))/\Gamma(d^{-1}(i+j)).$$

This and $a_{\mathbb{C}}(G,\omega)=1$ imply that the b-functions $b_i(s)$ of (G_i,X_i) are related as

$$b_2(s)/b_1(s) = \prod_{p<i\leq q} \prod_{0\leq j<d} (s+d^{-1}(i+j));$$

this is Shintani's formula; cf. [15], p. 78. We recall that if the roots $-\lambda$ of the b-function of (G,X) are all in $\mathbb{Z}$, then $Z(s)$ is known; in the p-adic case

$$Z(s) = \prod_{\lambda} (1-q_K^{-\lambda})/(1-q_K^{-(s+\lambda)});$$

cf. [10]. Therefore $Z(s)$ is "known" for the 8 types in § 5. We might further mention that (*) implies an explicit relation between $a_K(G_1,\omega_s)$ and $a_K(G_2,\omega_s)$ if $\ell=1$. Even if $\ell>1$ and ω is arbitrary in $\Omega(K^\times)$, it appears that $a_K(G_1,\omega)$ and $a_K(G_2,\omega)$ differ by a scalar factor provided that the orderings of $(G_i)_K$-orbits are compatible with the K-isomorphism of the Grassmannians. And finally if we start from a number field k as F, the product of the scalar factors for all $K=k_v$ is 1 provided that the product or products are taken separately expressing p-adic factors as quotients.

7. Zeta distributions $Z_a(\omega)$ and $Z_m(\omega)$.

We shall introduce two kinds of zeta distributions, one of which is a global complex power. We shall start by recalling some basic definitions refering the details to Serre [20], Tate [22] and Weil [23]. As before we shall denote by k a number field and use v instead of $K=k_v$ as subscript; for instance we shall write $|a_v|_v$ instead of $|a_v|_K$ if a_v is in K. We shall denote by Σ the set of all k_v or rather the set of all v and by Σ_∞ (resp. Σ_f) its subset defined by k_v being $\mathbb{R}$ or $\mathbb{C}$ (resp. a p-adic field). We shall also use the subscript v (resp. A) to denote the taking of k_v-rational points (resp. the adelization relative to k). If $a=(a_v)_v$ is an element of $k_A^\times$, the identity component of $\Omega(k_A^\times/k^\times)$ consists of

$$\omega_s(a)= |a|_A^s, \quad |a|_A= \prod_{v\in\Sigma} |a_v|_v$$

for all s in $\mathbb{C}$; and for ω arbitrary in $\Omega(k_A^\times/k^\times)$ a unique $\sigma(\omega)$ in $\mathbb{R}$ and for σ in $\mathbb{R}$ an open subset $\Omega_\sigma(k_A^\times/k^\times)$ of $\Omega(k_A^\times/k^\times)$ are defined as in the local case.

We take a regular prehomogeneous k-structure (G,X) in $X= \mathrm{Aff}^n$ with G irreducible and put

$$Z_a(\omega)(\phi)= \int_{G_A/G_k} \Big(\sum_{\xi\in Y_k} \phi(g\xi)\Big)\omega(\nu(g))\mu_G(g),$$

$$Z_m(\omega)(\phi)= \int_{Y_A} \phi(x)\omega(f(x))\mu_Y(x),$$

in which ϕ is arbitrary in $\mathcal{S}(X_A)$ and μ_G, μ_Y are the Tamagawa measures on G_A, Y_A. We have defined μ_Y regarding Y as a G-homogeneous space, hence μ_Y is a G_A-invariant measure on Y_A. In the special case where $(G,X)= (GL_1,\mathrm{Aff}^1)$ and $f(x)=x$, if we use the standard convergence factor $(\lambda_v)_v$, where $\lambda_v=1$ for v in Σ_∞ and $\lambda_v= (1-q_v^{-1})^{-1}$ for v in Σ_f, to define $\mu_G=\mu_Y$, then $Z_a(\omega)= Z_m(\omega)$ becomes the classical zeta distribution, say $Z_k(\omega)$, of k. In the general case the same $(\lambda_v)_v$ can be used to define μ_G, μ_Y. Furthermore by Ono [16] the integral $Z_m(\omega)(\phi)$ is absolutely convergent for every ϕ if $\sigma(\omega)>\kappa$, where $\kappa=n/d$ for $d= \deg(f)$. Therefore $Z_m(\omega)$ becomes an $\mathcal{S}(X_A)'$-valued holomorphic function on

$\Omega_\kappa(k_A^\times/k^\times)$ and, in fact, it becomes an element of $\mathcal{E}(\omega\circ\nu)=\mathcal{E}_{\mathfrak{X}}(\omega\circ\nu)$ for $\mathfrak{X}=X_A$. We recall that Weil's problem in this case is to determine all elements of $\mathcal{E}(\omega\circ\nu)$.

We might as well recall its solution in the case where $(G,X)=(GL_1,\mathrm{Aff}^1)$: If Θ denotes the subgroup of $k_A^\times$ consisting of all $a=(a_v)_v$ with a_v equal to the same positive real number for v in Σ_∞ and to 1 for v in Σ_f, then every ω in $\Omega(k_A^\times/k^\times)$ can be expressed uniquely as $\omega=\omega_s\chi$ with a character χ of $k_A^\times/k^\times$ satisfying $\chi|\Theta=1$. We take $\omega'=\omega_{s_o}\chi_o$ arbitrarily from $\Omega(k_A^\times/k^\times)$ and expand $Z_k(\omega)$ at $s=s_o$ in a Laurent series

$$Z_k(\omega)=T_o\cdot(s-s_o)^{e_o}+\ldots$$

with $T_o\neq0$; then $\mathcal{E}(\omega')=\mathbb{C}T_o$. In particular $\mathcal{E}(\omega')$ is 1-dimensional. As for the exponent e_o, it is 0 except for the following cases : If $\chi_o=1$ and $s_o=0,1$, then $e_o=-1$; if s_o is a zero of the Hecke zeta function $L(s,\chi_o)$ in the critical strip $0<\mathrm{Re}(s)<1$, then e_o is the order of that zero.

Now in the general case if the integral $Z_a(\omega)(\phi)$ is absolutely convergent for every ϕ, then $Z_a(\omega)$ also becomes an element of $\mathcal{E}(\omega\circ\nu)$. Therefore, as a preliminary to Weil's problem, it seems natural to ask how $Z_a(\omega)$ and $Z_m(\omega)$ are related.

8. Relation of $Z_a(\omega)$ and $Z_m(\omega)$.

Although $\mathcal{E}(\omega)$ is 1-dimensional, in the general case $\mathcal{E}(\omega\circ\nu)$ can be infinite dimensional. On the other hand G acts transitively on Y and G_v has a finite number of orbits in Y_v for every v. However G_A can have infinitely many orbits in Y_A. In fact we can show that

$$|G_A\backslash Y_A|=\prod_{v\in\Sigma}|G_v\backslash Y_v|;$$

hence the number of G_A-orbits in Y_A is finite if and only if G_v is transitive on Y_v for almost all v. This is the case if $\mathcal{E}(\omega\circ\nu)$ is finite dimensional for at least one ω in $\Omega_\kappa(k_A^\times/k^\times)$.

Now we have found that if the number of G_A-orbits in Y_A is finite, then $Z_a(\omega)$ also becomes an $\mathcal{S}(X_A)'$-valued holomorphic function on $\Omega_\kappa(k_A^\times/k^\times)$, the fixer G_ξ in G of any ξ in Y_k is connected semi-simple with the same Tamagawa number $\tau=\tau(G_\xi)$ and the identity

$$Z_a(\omega)=\tau.Z_m(\omega)$$

holds. This theorem has been proved in [12] under the following restriction : If the reduced pair in the class of (G,X) over an algebraic closure of k is of the form $(\rho(G_o\times GL_d),M_d)$, in which G_o is a connected irreducible subgroup of SL_d, then $G_o=SL_d$, i.e., $\rho(G_o\times GL_d)$ is the identity component of $Sim(det)$. It is very likely that the theorem is true without this restriction. At any rate if ω_v denotes the v-component of ω, by the results in § 6 we know the local factor

$$\int_{Y_v}\Phi_v(x)\omega_v(f(x)).\lambda_v|f(x)|_v^{-\kappa}\,dx_v$$

of $Z_m(\omega)(\Phi)$ if Φ_v is the characteristic function of $X(O_v)=O_v^n$, ω is unramified at v and G_v splits over k_v except when (G,X) is in the class of $(\Lambda^3GL_7,\Lambda^3Aff^7)$ over an algebraic closure of k.

We shall consider (G,X), in which G is the identity component of $Sim(f)$ for a nondegenerate quadratic form $f(x)$ in $n\geq 3$ variables with coefficients in k. If $D(f)$ is its discriminant, the number of G_A-orbits in Y_A is finite if and only if n is even and $D(f)$ is a square in $k^\times$. If this is not the case, then $Z_a(\omega)$ and $Z_m(\omega)$ are independent. More precise statement is as follows :

In general assume that there exists $h=\pm{}^th$ in $GL_n(k)$ satisfying $hGh^{-1}={}^tG$; choose a nontrivial character ψ of k_A/k and define Fourier transformations in $\mathcal{S}(X_A)$, $\mathcal{S}(X_A)'$ as in the local case via the bicharacter $\psi(tr\,{}^txhy)$ of X_A relative to the selfdual measure on X_A, i.e., its Haar measure normalized as $vol(X_A/X_k)=1$. Assume further that $Z_a(\omega)$ has a meromorphic continuation to the whole $\Omega(k_A^\times/k^\times)$ and satisfies the usual functional equation $Z_a(\omega)^*=Z_a(\omega_\kappa\omega^{-1})$. Assume finally that there exists at least one v such that G_v is not transitive on Y_v and the sums of entries of the columns of the Γ-matrix defined via the bicharacter $\psi_v({}^txhy)$ of X_v

are not all equal. Then there is no nonempty open subset of $\Omega_{\kappa}(k_A^{\times}/k^{\times})$ on which $Z_a(\omega)= c(\omega)Z_m(\omega)$ with a $\mathbb{C}^{\times}$-valued function $c(\omega)$.

If $f(x)$ is a quadratic form, we can use the h in $f(x)= 1/2.{}^t xhx$ as above. We recall that for $n>4$ the functional equation of $Z_a(\omega)$ is classical. In the case where n is odd or n is even and $D(f)$ is not a square in $k^{\times}$, we choose v from Σ_f such that ω is unramified at v, q_v is odd, the coefficients of $f(x)$ are in O_v and $D(f)$ is in $U_v= O_v-\pi_v O_v$ but not a square in U_v for n even. Then the condition on the Γ-matrix is also satisfied; cf. [9]. Therefore $Z_a(\omega)$ and $Z_m(\omega)$ are unrelated in the sense explained above.

We might mention that if n is even and $D(f)$ is a square in $k^{\times}$, then necessarily the sums of entries of the columns of the Γ-matrix are all equal for every v. This is in agreement with the following result in Gel'fand and Shilov [7], p. 284 : If $f(x)$ is an indefinite quadratic form of signature (p,q) with $p\neq q$ and $\omega= \omega_s(\mathrm{sgn})^m$, up to a scalar factor and relative to a suitable ordering of $G_{\mathbb{R}}$-orbits the Γ-matrix is given by

$$\begin{bmatrix} -\sin(s-p/2)\pi & (-1)^m\sin(p\pi/2) \\ (-1)^m\sin(q\pi/2) & -\sin(s-q/2)\pi \end{bmatrix} .$$

Therefore the sums of entries of the two columns are equal if and only if $p\equiv q \bmod 4$, i.e., if and only if n is even and $D(f)$ is a square in $\mathbb{R}^{\times}$.

We might further mention that the above criterion of independence can also be applied to the case where $f(x)$ is the discriminant of a binary cubic form, i.e., the case (S^3GL_2, S^3Aff^2). In that case the conditions are satisfied by Wright [25] and Datskovsky [4]; if k is not totally imaginary, we can even use Shintani's result in [21].

9. <u>Zeros and poles of</u> $Z_m(\omega)$.

We recall that the global complex power $Z_m(\omega)$ is holomorphic on $\Omega_{\kappa}(k_A^{\times}/k^{\times})$. However $Z_m(\omega)$ may not have a meromorphic continuation to the whole $\Omega(k_A^{\times}/k^{\times})$. On its domain of meromorphy, which is the largest open subset of $\Omega(k_A^{\times}/k^{\times})$ where $Z_m(\omega)$ is meromorphic, it can certainly have zeros. In fact in the simplest case where $f(x)=x$ the zeros of $Z_m(\omega)= Z_k(\omega)$ are precisely the zeros of $L(s,\chi)$ in the critical strip $0<\mathrm{Re}(s)<1$. Therefore the determination of actual zeros and poles of $Z_m(\omega)$ is more than the Riemann hypothesis even in this case. We shall denote by $R(\chi)$ the formal series $\Sigma(\rho)$, in which ρ runs over the set of all zeros of $L(s,\chi)$ in the critical strip each repeated with its multiplicity. Then the divisor of $Z_m(\omega)$ in its domain of meromorphy can be expressed in terms of the translate $R(\chi)_\sigma$ of $R(\chi)$ under $s \longmapsto s+\sigma$ in the case where $f(x)$ is a quadratic form. This is based on the fact that the actual poles on $\Omega(K^\times)$ of the complex power of a quadratic form $f(x)$ can be determined for all $K=k_v$. At any rate, refering the details to [12], we shall explain some results :

In the case of a quadratic form $f(x)$ the domain of meromorphy of $Z_m(\omega)$ is $\Omega(k_A^\times/k^\times)$ if n is even while it is the complement of the union of all lines $\mathrm{Re}(s)=-\kappa$ in the connected components of $\Omega(k_A^\times/k^\times)$ if n is odd. Furthermore if we assume, for the sake of simplicity, that n is even, $n\geq 6$ and $D(f)$ is a square in $k^\times$, the divisor of $Z_m(\omega_s)$ becomes

$$R(1)+R(1)_{\kappa-1}-\{(0)+(1)+(\kappa-1)+(\kappa)\}+\sum_{f_v \text{ definite}}\ \sum_{0\leq j<\kappa/2}(\kappa-2j-1),$$

in which f_v is f over $k_v=\mathbb{R}$. Therefore if $k=\mathbb{Q}$, $n\equiv 0 \bmod 4$ and $f(x)$ is positive-definite, then $Z_m(\omega_s)$ and $\Gamma(s)\zeta(s)\zeta(s-\kappa+1)$ have the same divisor. If further $f(x)$ is $\mathbb{Z}$-valued on $\mathbb{Z}^n$ and $D(f)=1$, then the identity $\tau^{-1}.Z_a(\omega_s)(\phi)= Z_m(\omega_s)(\phi)$ specializes to the following classical formula :

$$\int_0^\infty\{\Sigma(cyi+d)^{-\kappa}\}y^{s-1}dy= (\Gamma(\kappa)\zeta(\kappa))^{-1}(2\pi)^{\kappa-s}\Gamma(s)$$
$$.\prod_p(1-(1+p^{\kappa-1})p^{-s}+p^{\kappa-1-2s})^{-1},$$

in which the summation on the L.H.S. is extended over the set of relatively prime integer pairs (c,d) with $c>0$. The special ϕ which gives the above formula is $\otimes_p \phi_p \otimes \phi_\infty$, where ϕ_p is the characteristic function of $\mathbb{Z}_p^n$ for every p and $\phi_\infty(x)= \exp(-2\pi f(x))$. We have tacitly used the Siegel formula. If we do not use that formula, the integrand on the L.H.S. becomes a weighted mean of theta series associated with classes in the genus of $f(x)$.

(*) p. 67 : This work was partially supported by the National Science Foundation.

BIBLIOGRAPHY

[1] M.F. Atiyah.- Resolution of singularities and division of distributions, Comm. pure and appl. math. 23 (1970), 145-150.

[2] I.N. Bernshtein and S.I. Gel'fand.- Meromorphic property of the functions P^{λ}, Functional Analysis and its Applications 3 (1969), 68-69.

[3] I.N. Bernshtein.- The analytic continuation of generalized functions with respect to a parameter, Functional Analysis and its Applications 6 (1972), 273-285.

[4] B.A. Datskovsky.- On zeta functions associated with the space of binary cubic forms with coefficients in a function field, Thesis, Harvard, 1984.

[5] J. Denef.- The rationality of the Poincaré series associated to the p-adic points on a variety, Invent. Math. 77 (1984), 1-23.

[6] J. Denef.- On the degree of Igusa's local zeta function, preprint.

[7] I.M. Gel'fand and G.E. Shilov.- Generalized functions. I, Academic Press (1964).

[8] J. Igusa.- Complex powers and asymptotic expansions. I. Crelles J. Math. 268/269 (1974), 110-130; II. Ibid. 278-279 (1975), 307-321.

[9] J. Igusa.- Some results on p-adic complex powers, Amer. J. Math. 106 (1984), 1013-1032.

[10] J. Igusa.- On functional equations of complex powers, Invent. math. 85 (1986), 1-29.

[11] J. Igusa.- On a certain class of prehomogeneous vector spaces, to appear in J. of Algebra.

[12] J. Igusa.- Zeta distributions associated with some invariants, to appear in Amer. J. Math.

[13] V.G. Kac, V.L. Popov and E.B. Vinberg.- Sur les groupes linéaires algébriques dont l'algèbre des invariants est libres, C. R. Acad. Sc. Paris 283 (1976), 875-878.

[14] M. Kashiwara.- B-functions and holonomic systems (Rationality of b-functions), Invent. Math. 38 (1976), 33-53.

[15] T. Kimura.- The b-functions and holonomy diagrams of irreducible regular prehomogeneous vector spaces, Nagoya Math. J. 85 (1982), 1-80.

[16] T. Ono.- An integral attached to a hypersurface, Amer. J. Math. 90 (1968), 1224-1236.

[17] M. Sato and T. Kimura.- A classification of irreducible prehomogeneous vector spaces and their relative invariants, Nagoya Math. J. 65 (1977), 1-155.

[18] M. Sato and T. Shintani.- On zeta functions associated with prehomogeneous vector spaces, Ann. Math. 100 (1974), 131-170.

[19] J.-P. Serre.- Cohomologie Galoisienne.- Lect. Notes in Math. 5, Springer-Verlag (1965).

[20] J.-P. Serre.- Méthodes adéliques.- Résumé des cours et travaux, Collège de France (1981-1982), 81-89.

[21] T. Shintani.- On Dirichlet series whose coefficients are class numbers of integral binary cubic forms, J. Math. Soc. Japan 24 (1972), 132-188.

[22] J. Tate.- Fourier analysis in number fields and Hecke's zeta-functions, Thesis, Princeton, 1950; Algebraic Number Theory, Acad. Press. (1967), 305-347.

[23] A. Weil.- Adeles and algebraic groups.- Institute for Advanced Study, Princeton (1961); Progress in Math. 23, Birkhäuser (1982).

[24] A. Weil.- Fonction zêta et distributions, Sém. Bourbaki 312 (1966), 1-9; Collected Papers III, Springer-Verlag (1980), 158-163.

[25] D.J. Wright.- The adelic zeta function associated to the space of binary cubic forms, Math. Ann. 270 (1985), 503-534.

J.-I. IGUSA
The Johns Hopkins University
Department of Mathematics
Baltimore, MD 21218
U.S.A.

SURVEY OF THE PROOF OF THE TATE CONJECTURES FOR HILBERT-BLUMENTHAL SURFACES

Christoph KLINGENBERG

In this lecture the main steps in the proof of the Tate conjectures for Hilbert-Blumenthal surfaces will be discussed. This conjecture was proved for most cases by Harder, Langlands and Rapoport in their fundamental paper [9]. The remaining cases are covered by my thesis [11]. Recently Ramakrishnan and Murty gave a different proof for these remaining cases in their paper [14]. For more information on Hilbert-Blumenthal surfaces, especially on the Beilinson conjectures, I recommend the excellent survey article of Ramakrishnan [16]. I would like to thank Professor G. Henniart very much for inviting me to this seminar.

1. The Tate Conjectures for Algebraic Surfaces Defined over $\mathbb{Q}$.

1.1. The Problem.

Let S be a smooth projective surface defined over $\mathbb{Q}$. The starting point is the question :

"How many" algebraic curves are on S ?

To be more precise, we look at two fundamental invariants associated with S, namely the Betti and the ℓ-adic cohomology and reformulate the notion of algebraic curve in cohomological terms by means of the cycle mapping c. This mapping c maps the free abelian group generated by the irreducible algebraic curves on S into the second cohomology, see [4], p. 20. We call a class in the image of c an algebraic cycle. The above question now reads :

Which cohomology classes are algebraic cycles ?

1.2. Betti Cohomology.

By the above hypotheses, $S(\mathbb{C})$ is a compact complex manifold, so we can consider its second Betti cohomology $H_B^2(S)$. This $\mathbb{Q}$-vector space has a Hodge decomposition over $\mathbb{C}$, see [7], Ch. 1 :

$$H_B^2(S) \otimes_{\mathbb{Q}} \mathbb{C} = H^{2,0}(S) \oplus H^{1,1}(S) \oplus H^{0,2}(S).$$

1.3. ℓ-adic Cohomology.

On the other hand for a prime number ℓ we consider the ℓ-adic cohomology group $H^2_\ell(S\times\bar{\mathbb{Q}})$ as a $\mathbb{Q}_\ell$-vector space with an action ρ of $Gal(\bar{\mathbb{Q}}/\mathbb{Q})$. We have a comparison isomorphism (see [4], p. 19) :

$$\Phi : H^2_B(S) \otimes_{\mathbb{Q}} \mathbb{Q}_\ell \longrightarrow H^2_\ell(S\times\bar{\mathbb{Q}}).$$

1.4. Rationality Questions.

In order to distinguish between two notions of rationality Deligne defines in his article [5] the field of coefficients and the field of definition of a motive. Using this terminology one can say that the field of definition of Betti cohomology (resp. ℓ-adic cohomology) is $\mathbb{C}$ (resp. $\bar{\mathbb{Q}}$), while its field of coefficients is $\mathbb{Q}$ (resp. $\mathbb{Q}_\ell$). For technical reasons one sometimes has to extend the field of coefficients, e.g. for the above comparison isomorphism or for the operation of the Hecke algebra, see section 3.

We call a cohomology class $\xi \in H^2_B(S) \otimes_{\mathbb{Q}} \mathbb{C}$ Betti rational, if $\xi \in H^2_B(S)$, or, in other words, if the field of cofficients of ξ is $\mathbb{Q}$.

The Galois representation ρ acts on the field of definition of the ℓ-adic cohomology.

1.5. The Conjectures.

In order to turn the above question into more precise conjectures, one simply analyses the properties of algebraic cycles in the various cohomology theories. For Betti cohomology this leads to the following definition :

Definition. A cohomology class $\xi \in H^2_B(S) \otimes_{\mathbb{Q}} \mathbb{C}$ is called a Hodge cycle, if

(i) $\xi \in H^2_B(S)$ that is, ξ is Betti rational, and

(ii) $\xi \in H^{1,1}(S)$.

The standard examples for Hodge cycles are the algebraic cycles, see [7], p. 163. One can, of course, generalize this notion to projective smooth varieties of any dimension. A Hodge cycle of codimension p then is a Betti rational class of type (p,p). The Hodge conjecture asserts :

All Hodge cycles on a projective smooth variety are algebraic cycles.

Fortunately this conjecture is proved in codimension one (see [7], p. 163), especially :

<u>Theorem</u> (Lefschetz). All Hodge cycles on $S(\mathbb{C})$ are algebraic.

The ℓ-adic version of the Hodge conjecture reads as follows : let $\mathbb{Q} \subset M \subset \overline{\mathbb{Q}}$ be a finite Galois extension and $\alpha : \mathrm{Gal}(\overline{\mathbb{Q}}/\mathbb{Q}) \longrightarrow \mathbb{Z}_\ell^*$ the Tate character.

<u>Definition</u> [20]. A cohomology class $\xi \in H_\ell^2(S\times\overline{\mathbb{Q}})$ is called a <u>Tate cycle defined over</u> M, if

$$\rho(\sigma)\xi = \alpha^{-1}(\sigma)\xi \quad \text{for all } \sigma \in \mathrm{Gal}(\overline{\mathbb{Q}}/M).$$

Again the standard examples are the algebraic cycles defined over M, see [20], § 2. Using Deligne's terminology a Tate cycle over M is an ℓ-adic cohomology class with M as a field of definition.

The <u>Tate conjectures over</u> M assert :

1. All Tate cycles defined over M are algebraic.
2. The rank of the subspace of algebraic cycles defined over M equals the order of pole at $s=2$ of the Hasse-Weil-Zetafunction of S over M.

<u>Remark</u> :

When we pass to a bigger M, we get more Tate cycles defined over M and the Tate conjectures become more difficult to prove.

2. <u>Hilbert-Blumenthal Surfaces</u>.
2.1. <u>The Definition</u>.

Let F be real quadratic, $\mathcal{O}_F \subset F$ the ring of integers in F, $A_\mathbb{Q} = \mathbb{R}\times A_f$ (resp. A_F) the adeles over $\mathbb{Q}$ (resp. F) and $G := R_{F/\mathbb{Q}}GL_2$, so we have $G(\mathbb{Q}) = GL_2(F)$ and $G(A_\mathbb{Q}) = GL_2(A_F)$. Choose

$$K_\infty = \left\{\begin{bmatrix} a_1 & -b_1 \\ b_1 & a_1 \end{bmatrix} \times \begin{bmatrix} a_2 & -b_2 \\ b_2 & a_2 \end{bmatrix}\right\} \subset G(\mathbb{R}) \simeq GL_2(\mathbb{R}) \times GL_2(\mathbb{R}),$$

then we have $K_\infty = [Z^0.SO(2)]\times[Z^0.SO(2)]$ with $Z^0 \simeq \mathbb{R}_{>0}$. An open compact subgroup $K_f \subset G(A_f)$ determines a Shimura variety S_{K_f}, the complex points of which are given by

$$\begin{aligned} S_{K_f}(\mathbb{C}) &= G(\mathbb{Q})\backslash G(A)/K_\infty.K_f \\ &= G(\mathbb{Q})\backslash[G(\mathbb{R})/K_\infty \times G(A_f)/K_f]. \end{aligned}$$

For $K'_f \subset K_f$ we have a finite morphism $S_{K'_f} \longrightarrow S_{K_f}$. So we can define the inverse limit :

$$S := \varprojlim_{K_f} S_{K_f} = G(\mathbb{Q})\backslash G(A)/K_\infty .$$

<u>Definition</u>. We call S the <u>Hilbert-Blumenthal surface</u> (or <u>Hilbert modular surface</u>) associated to F.

The scheme S is quasi-compact and separated, but not of finite type over $\mathbb{C}$, cf. [3], 1.8.2 or [2], 2.1.4. Rapoport constructs in [18] a proper regular model for S_{K_f} over $\mathbb{Q}$.

2.2. <u>Connected Components</u>.

By repeating the above process for the algebraic group $\tilde{G} := R_{F/\mathbb{Q}}SL_2$, we get a connected component $\tilde{S}$ of S. The connected components are precisely the fibers of the determinant mapping :

$$\det : S(\mathbb{C}) \longrightarrow I_F/(F^*.(\mathbb{R}^2_{>0}\times\{1\})).$$

The variety $\tilde{S}$ corresponds to the "kernel" of this mapping. So think of the connected components of S as of the layers isomorphic to SL_2 inside GL_2. The surface $\tilde{S}$ classifies abelian surfaces with real multiplication, a fixed Tate module and a fixed value of the Weil pairing. According to a result due to Shimura [19], the surface $\tilde{S}$ is defined over $\mathbb{Q}^{ab}$, the maximal abelian extension of $\mathbb{Q}$. The union S of the various connected components is defined over $\mathbb{Q}$. The crucial idea in the proof later on is to investigate how the cohomology of S relates to the cohomology of its connected components.

2.3. <u>Cuspidal Cohomology</u>.

In the following we consider the cuspidal cohomology of S :

$$H^2_{cusp}(S) := \operatorname{Im}(H^2_c(S) \longrightarrow H^2(S)).$$

This part of the cohomology has similar properties as the cohomology of a smooth projective surface, cf. [15], Ch. 1, or [8]. Harder, Langlands and Rapoport ([9], 2.1) indicate how to modify the Tate conjectures for a non-compact surface.

The first step in the proof of the Tate conjectures is to identify the Tate cycles. To be able to deal with the infinite-dimensional Galois representation ρ, we chop the cohomology into palatable pieces by means of automorphic forms.

In the following we shall exclusively deal with the cuspidal part of the cohomology, so we may omit the subscript "cusp".

3. Automorphic Forms.

3.1. Decomposition of the Betti Cohomology.

After extension of the field of definition the Hecke algebra $G(A_f)$ acts on $H^2_B(S) \otimes_{\mathbb{Q}} \mathbb{C}$, so we can decompose the (cuspidal) cohomology as follows. Let $\pi = \pi_\infty \otimes \pi_f$ be a cuspidal automorphic form for G realized on a space $H_\pi = H_{\pi_\infty} \otimes H_{\pi_f}$. Then we have a decomposition

$$H^2_B(S) \otimes_{\mathbb{Q}} \mathbb{C} = \bigoplus_{\pi_f} H^2_B(S)(\pi_f) \quad \text{with}$$

$$\begin{aligned} H^2_B(S)(\pi_f) &= \mathrm{Hom}_{G(A_f)}(H_{\pi_f}, H^2_B(S) \otimes_{\mathbb{Q}} \mathbb{C}) \otimes H_{\pi_f} \\ &=: H^2_{B,\infty}(S)(\pi_f) \otimes H_{\pi_f}. \end{aligned}$$

Remark :

The above direct sum is actually a proper subspace of the cuspidal cohomology, because some one-dimensional representations π contribute to the direct sum, see [9], 1.8. But we shall concentrate on the interesting part of the cohomology and continue to assume π to be cuspidal automorphic.

3.2. Computation of the Betti Cohomology.

We calculate an isotypical component of the Betti cohomology of S by means of g-K_∞-cohomology. Let g be the Lie algebra of G and $V(\pi_\infty)$ the space of C^∞-vectors in the representation space of π_∞, then

$$H^2_{B,\infty}(S)(\pi_f) \simeq H^2(g, K_\infty, V(\pi_\infty)).$$

By the standard vanishing theorem of the g-K_∞-cohomology [1], I, Th. 5.3, only those automorphic forms π contribute to the cohomology where π_∞ is the first discrete series representation of $G(\mathbb{R})$. Denote by Z the center of $GL_2(\mathbb{R})$ and by $gl(2)$, $so(2)$ and z the Lie algebras of $GL_2(\mathbb{R})$, $SO(2)$ and Z. We decompose $\pi_\infty = \pi_{\infty,1} \otimes \pi_{\infty,2}$, $V(\pi_\infty) = V(\pi_{\infty,1}) \otimes V(\pi_{\infty,2})$ and get from the Künneth formula [1], I, 1.3. :

$$\begin{aligned}
H^2(g,K_\infty,V(\pi_\infty)) &= H^1(gl(2),Z.SO(2),V(\pi_{\infty,1})) \\
&\quad \otimes H^1(gl(2),Z.SO(2),V(\pi_{\infty,2})) \\
&= \mathrm{Hom}_{Z.SO(2)}(gl(2)/(z.so(2)),V(\pi_{\infty,1})) \\
&= \otimes\ \mathrm{Hom}_{Z.SO(2)}(gl(2)/(z.so(2)),V(\pi_{\infty,2})).
\end{aligned}$$

<u>Proposition</u>. For a cuspidal automorphic form $\pi = \pi_\infty \otimes \pi_f$ the space

$$H^2_{B,\infty}(S)(\pi_f) = H^{2,0}(S)(\pi_f) \oplus H^{1,1}(S)(\pi_f) \oplus H^{0,2}(S)(\pi_f)$$

is a four-dimensional vector space over $\mathbb{C}$ such that the (2,0)-term and the (0,2)-term are one-dimensional and the (1,1)-term is two-dimensional.

This follows from the fact that, for example, the space $H^{2,0}(S)(\pi_f)$ consists of homomorphisms mapping the matrix $\begin{pmatrix}1 & i\\ i & -1\end{pmatrix}$ to the holomorphic vector of lowest weight on both archimedean places of F, see [11], 2.6.

3.3. <u>Betti rationality</u>.

Consider the action of $\mathrm{Aut}(\mathbb{C}/\mathbb{Q})$ on the field of coefficients, that is, on the second factor of $H^2_B(S) \otimes_{\mathbb{Q}} \mathbb{C}$. This action factorizes over the finite-dimensional spaces $H^2_B(S_{K_f}) \otimes_{\mathbb{Q}} \mathbb{C}$, so for every π_f there exists a finite extension $\mathbb{Q}(\pi_f)$ of $\mathbb{Q}$ such that $\mathrm{Aut}(\mathbb{C}/\mathbb{Q}(\pi_f))$ leaves invariant the space $H^2_B(S)(\pi_f)$. In order to keep the notation simple we shall assume that $\mathbb{Q}(\pi_f)=\mathbb{Q}$. In the general case one has to consider simultaneously the $\mathrm{Aut}(\mathbb{C}/\mathbb{Q})$-orbit of this space, see [11], 2.8.

3.4. <u>Decomposition of the ℓ-adic cohomology</u>.

The action of the Hecke algebra also gives a decomposition

$$H^2_\ell(S\times\overline{\mathbb{Q}}) \otimes_{\mathbb{Q}_\ell} \overline{\mathbb{Q}}_\ell = \bigoplus_{\pi_f} H^2_{\ell,\infty}(S\times\overline{\mathbb{Q}})(\pi_f) \otimes H_{\pi_f},$$

where $H^2_{\ell,\infty}(S\times\overline{\mathbb{Q}})(\pi_f)$ is a four-dimensional $\overline{\mathbb{Q}}_\ell$-vector space. The $G(A_f)$-action commutes with the Galois action on the ℓ-adic cohomology, so we get four-dimensional representations

$$\rho(\pi_f) : \mathrm{Gal}(\overline{\mathbb{Q}}/\mathbb{Q}) \longrightarrow GL(H^2_{\ell,\infty}(S\times\overline{\mathbb{Q}})(\pi_f)).$$

We are led to the following problems :

Problem A : For which automorphic forms π do Tate cycles exist in $H^2_{\ell,\infty}(S\times\bar{\mathbb{Q}})(\pi_f)$?

Problem B : Are they algebraic ?

4. Identificatioon of the Tate Cycles.

In order to formulate the corresponding Theorem A, we introduce two types of automorphic forms. For GL_2 there exists a notion of lifting for automorphic forms. In the special case of lifting from $\mathbb{Q}$ to F this was first presented explicitly by Doi and Naganuma [6]; later Langlands (after Saito and Shintani) more generally investigated the case of cyclic extension of prime degree, see [13]. In classical terms the lifting from $\mathbb{Q}$ to F associates a Hilbert modular form with an elliptic modular form. We call an automorphic form distinguished, if it is a tensor product of a lifted automorphic form and a certain character, see [9], Def. 2.7 for the precise definition.

For the other special class of automorphic forms fix a quadratic character

$$\omega : I_F/F^* \longrightarrow \{\pm 1\}$$

and call π an automorphic form of CM-type, if $\pi \simeq \pi \otimes \omega$. The character ω defines a quadratic extension E of F and one can show that π is "induced" (in a sense explained in [13], esp. Lemma 7.17) from a Hecke character θ on I_E/E^*, write $\pi = \pi(\theta)$.

Theorem A (Harder, Langlands and Rapoport [9]).

(i) For M over $\mathbb{Q}$ abelian there exist Tate cycles only for distinguished automorphic forms.
(ii) For M over $\mathbb{Q}$ non-abelian there exist Tate cycles which are not defined over any abelian extension of $\mathbb{Q}$ only for automorphic forms of CM-type. These cycles form a two-dimensional subspace in the four-dimensional $H^2_{\ell,\infty}(S\times\bar{\mathbb{Q}})(\pi_f)$.

Let me briefly explain this result. Oda [15] observed that the Galois action $\rho(\pi)$ leaves invariant the intersection form on the second cohomology, so the elements of $Gal(\bar{\mathbb{Q}}/\mathbb{Q})$ act as orthogonal transformations. According to the decomposition of the group of orthogonal transformations, one can decompose $\rho(\pi)$ over F into the product of two two-dimensional representations :

$$\rho(\pi)\,|Gal(\bar{\mathbb{Q}}/F) = \rho_1(\pi) \otimes \rho_2(\pi).$$

Of course the $\rho_i(\pi)$ are not uniquely determined. The proof of Theorem A now consists of a careful analysis of the invariant subspaces of the two-dimensional representations $\rho_i(\pi)$. But one can say a little more about the $\rho_i(\pi)$. According to a well-known conjecture of

Langlands there should correspond to an automorphic form π over F a two-dimensional representation $\tau(\pi)$ of $Gal(\bar{\mathbb{Q}}/F)$, so that the local factors of the automorphic L-function and the Galois L-function coincide for almost all places of F. In the two special cases above we can explicitely determine $\tau(\pi)$: for a distinguished automorphic form π the representation $\tau(\pi)$ is given by the Eichler-Shimura theory and for $\pi = \pi(\theta)$ of CM-type we have

$$\tau(\pi) = \mathrm{Ind}_{Gal(\bar{\mathbb{Q}}/E)}^{Gal(\bar{\mathbb{Q}}/F)} \theta,$$

where θ is the above mentioned Hecke character of I_E/E^* which we consider as a $Gal(\bar{\mathbb{Q}}/E)$-character by class-field theory.

In the special case of a distinguished or a CM-type automorphic form one can choose $\rho_i(\pi)$, such that $\tau(\pi) \simeq \rho_1(\pi)$ and $\tau(\pi)^\sigma \simeq \rho_2(\pi)$ for the non-trivial element $\sigma \in Gal(F/\mathbb{Q})$, see [9], Kap. 4. For an arbitrary automorphic form one expects the same equalities to hold, but there seems to be no proof in sight.

5. The Proof.
5.1. Abelian Extensions.

Theorem B 1 (Harder, Langlands and Rapoport [9]).
All Tate cycles defined over an abelian extension of $\mathbb{Q}$ are algebraic.

Hirzebruch and Zagier [10] invented a method to construct cycles on a Hilbert-Blumenthal surface. Think of G as the product of two copies of GL_2 and embed diagonally the Shimura variety associated to $GL_2/\mathbb{Q}$. This gives an embedding of modular curves into the surface. The modular interpretation for these modular curves is given by those abelian surfaces, which are products of two elliptic curves over $\mathbb{Q}$. The Hecke operators translate these curves into the various parts of cohomology and it is intuitively clear that these cycles exhaust the Tate cycles in the components $H^2_\ell(S \times \bar{\mathbb{Q}})(\pi_f)$ corresponding to distinguished forms π.

To prove the Tate conjectures over non-abelian extensions of $\mathbb{Q}$ we use a special property of CM-forms, their instability, to obtain a two-dimensional subspace of Hodge cycles in the corresponding $H^2_B(S)(\pi_f)$. By Lefschetz's theorem these cycles are algebraic. It would be very profitable, however, to construct these cycles explicitly and to give a modular interpretation of them, if possible. With a method to construct these cycles one oculd attack the Beilinson conjectures over non-abelian extensions of $\mathbb{Q}$, see Ramakrishnan's paper [17] for the abelian case.

5.2. Non-abelian Extensions.

In the rest of this lecture I would like to indicate a proof of the following

<u>Theorem B 2</u> [11]. Over a non-abelian extension $M \supset \mathbb{Q}$ all Tate cycles are Hodge cycles and therefore algebraic.

According to Theorem A (ii) we have to consider a CM-form π, which gives a two-dimensional subspace of Tate cycles in the four-dimensional $H^2_{\ell,\infty}(S\times\bar{\mathbb{Q}})(\pi_f)$. On the other hand the subspace $H^{1,1}(S)(\pi_f)$ in $H^2_{B,\infty}(S)(\pi_f)$ also is two-dimensional, so we have to prove that this subspace is Betti rational. The "slopes" of this subspace are given by the ratios of the periods of π. One could thus prove the rationality of $H^{1,1}(S)(\pi_f)$ by establishing the corresponding period relation. Ramakrishnan and Murty (see introduction) have taken this path. We prove that for a connected component $\tilde{S}$ of S the space $H^{1,1}(\tilde{S})(\pi_f)$ either vanishes or equals the full $H^2_{B,\infty}(\tilde{S})(\pi_f)$, so that the Betti rationality follows trivially in both cases. In order to be more precise (and to give a meaning to the symbol $H^2_{B,\infty}(\tilde{S})(\pi_f)$) we have to restrict automorphic forms on $G = R_{F/\mathbb{Q}}GL_2$ to $\tilde{G} = R_{F/\mathbb{Q}}SL_2$. Here for CM-forms the following surprising phenomenon occurs.

5.3. <u>Instability of Automorphic Forms</u>.

Let π be an arbitrary form on $G(A)$ and let Π be the set of irreducible components of the restriction of π to $\tilde{G}(A)$. The set Π is called the <u>L-packet</u> of π.

<u>Definition</u>. The automorphic form π is called <u>instable</u>, if the elements of Π do not occur with the same multiplicity in the space of cusp forms on $\tilde{G}(A)$.

<u>Proposition</u> (Labesse and Langlands [12]). CM-forms are instable. More precisely, there exists a partition $\Pi = \Pi^+ \cup \Pi^-$ such that the elements $\tilde{\pi}\in\Pi^+$ are automorphic on $\tilde{G}(A)$ and the elements $\tilde{\pi}\in\Pi^-$ are not.

Let us motivate this result, see [11], Kap. 3 for details. The quadratic extension E of F associated with a CM-form π defines via the norm mapping a subgroup $G^+(A) \subset G(A)$ such that $G^+(A).G(\mathbb{Q})$ has index 2 in $G(A)$. Think of π and $\pi\otimes\omega$ as different automorphic forms with different Whittaker models $\mathcal{W}(\pi,\tau)$ and $\mathcal{W}(\pi\otimes\omega,\tau)$ for an additive character τ of A_F/F. The isomorphism $\pi \simeq \pi\otimes\omega$ forces $\mathcal{W}(\pi,\tau) \simeq \mathcal{W}(\pi\otimes\omega,\tau)$. The only way to establish this isomorphism between the "different" spaces is a decomposition

$$\mathcal{W}(\pi,\tau) = \mathcal{W}^+(\pi,\tau) \oplus \mathcal{W}^-(\pi,\tau)$$

of the Whittaker model into subspaces of functions with support in $G^+(A).G(\mathbb{Q})$ and support outside $G^+(A).G(\mathbb{Q})$, respectively. When we restrict the representation to $\tilde{G}(A)$ only the functions in $\mathcal{W}^{+}(\pi,\tau)$ will give representation spaces for automorphic forms on $\tilde{G}(A)$, the others will vanish. In other words, if you determine the Fourier expansion of an element $\tilde{\pi}\in\Pi^{-}$ on $\tilde{G}$, you will see that all Fourier coefficients vanish, so $\tilde{\pi}$ cannot be automorphic.

5.4. <u>The Cohomology of a Connected Component</u>.

Let $\tilde{S}\subset S$ be a connected component and $\tilde{\pi}=\tilde{\pi}_{\infty}\otimes\tilde{\pi}_{f}\in\Pi$ an element in the L-packet of a CM-automorphic form π. Denote by $\tilde{\pi}_{\infty}^{+}$ (resp. $\tilde{\pi}_{\infty}^{-}$) the holomorphic (resp. anti-holomorphic) discrete series representation of $SL_2(\mathbb{R})$.

<u>Proposition</u>. We have

$$H^{2}_{B,\infty}(\tilde{S})(\tilde{\pi}_{f}) \simeq \begin{cases} H^{2,0}(\tilde{S})(\tilde{\pi}_{f}) \oplus H^{0,2}(\tilde{S})(\tilde{\pi}_{f}) & \text{or} \\ H^{1,1}(\tilde{S})(\tilde{\pi}_{f}), \end{cases}$$

depending on whether $\tilde{\pi}_{\infty}^{+}\otimes\tilde{\pi}_{\infty}^{+}\otimes\tilde{\pi}_{f}$ and $\tilde{\pi}_{\infty}^{-}\otimes\tilde{\pi}_{\infty}^{-}\otimes\tilde{\pi}_{f}$ are automorphic on $\tilde{G}(A)$ or not.

This is a direct consequence of the instability of CM-forms. Namely, given the finite part π_f of an automorphic form on $G(A)$, because of Multiplicity One for G there is exactly one representation π_∞ of $G(\mathbb{R})$, so that $\pi=\pi_{\infty}\otimes\pi_{f}$. As

$$\pi_{\infty}|G(\mathbb{R}) = \tilde{\pi}_{\infty}^{+}\otimes\tilde{\pi}_{\infty}^{+} \oplus \tilde{\pi}_{\infty}^{+}\otimes\tilde{\pi}_{\infty}^{-} \oplus \tilde{\pi}_{\infty}^{-}\otimes\tilde{\pi}_{\infty}^{+} \oplus \tilde{\pi}_{\infty}^{-}\otimes\tilde{\pi}_{\infty}^{-},$$

we have four choices to complete the finite part $\tilde{\pi}_f$ of an automorphic form on $\tilde{G}(A)$ to a <u>representation</u> of $\tilde{G}(A)$. Because of the instability of CM-forms only two of these choices will give an <u>automorphic form</u> on $\tilde{G}(A)$. It follows from the description of the Hodge decomposition that

$$\tilde{\pi}_{\infty}^{+}\otimes\tilde{\pi}_{\infty}^{+}\otimes\tilde{\pi}_{f} \text{ is automorphic on } \tilde{G}(A) \iff H^{2,0}(\tilde{S})(\tilde{\pi}_{f}) \neq 0.$$

In the first case of the proposition the (1,1)-term vanishes and in the second case it equals the full second cohomology. So in either case we get :

<u>Corollary</u>. $H^{1,1}(\tilde{S})(\tilde{\pi}_f)$ is Betti-rational.

Collecting this information on the various connected components we get Theorem B2.

Let us summarize the proof. Take a connected component $\tilde{S} \subset S$ corresponding to a layer isomorphic to SL_2 inside GL_2, see 2.2. If we restrict the functions in $\mathcal{W}(\pi,\tau)$, the Whittaker model of a CM-automorphic form π, "half" of them will vanish on this layer, so only one half of the L-packet of π will give an automorphic form on this layer. This reduces the four-dimensional $H^2_{B,\infty}(S)(\pi_f)$ to a two-dimensional space $H^2_{B,\infty}(\tilde{S})(\tilde{\pi}_f)$ for a $\tilde{\pi}\in\Pi$. This space $H^2_{B,\infty}(\tilde{S})(\tilde{\pi}_f)$ can be recovered either as the $H^{2,0}(\tilde{S})(\tilde{\pi}_f) \oplus H^{0,2}(\tilde{S})(\tilde{\pi}_f)$ or as the $H^{1,1}(\tilde{S})(\tilde{\pi}_f)$. So in either case it follows that $H^{1,1}(\tilde{S})(\tilde{\pi}_f)$ is Betti-rational.

The second claim of the Tate conjectures follows from the main result of Harder, Langlands and Rapoport [9], Satz 2.4. They compare the motivic L-function with the automorphic L-function by Langlands' method of comparing the Selberg Trace Formula with the Lefschetz Fixed Point Formula.

BIBLIOGRAPHY

[1] A. Borel, N. Wallach.- Continuous Cohomology, Discrete Subgroups and Representations of Reductive Groups. PUP Annals of Mathematics Studies, N° 94.

[2] P. Deligne.- Varietés de Shimura, In : Automorphic Forms, Representations and L-Functions. Proc. of Symp. in Pure Math. vol 33, (1979) (Corvallis), II, p. 247-290.

[3] P. Deligne.- Travaux de Shimura. Séminaire Bourbaki, Exp. 389.

[4] P. Deligne, J. Milne, A. Ogus, K. Shih.- Hodge Cycles, Shimura Varieties and Motives. Lecture Notes in Math. N° 900, Springer, Heidelberg. We only refer to the first paper by P. Deligne : Hodge Cycles on Abelian Varieties.

[5] P. Deligne.- Valeurs de Fonctions L et Périodes d'Intégrales. In : Automorphic Forms, Representations and L-Functions. Proc. of Symp. in Pure Math. vol. 33, (1979) (Corvallis), II p. 313-346.

[6] K. Doi, H. Naganuma.- On the functional equation of certain Dirichlet series. Inv. Math. 9 (1969), p. 1-14.

[7] P. Griffiths, J. Harris.- Principles of Algebraic Geometry. John Wiley (1978).

[8] G. Harder.- The Cohomology of $SL_2(\mathcal{O})$. in : Lie Groups and their Representations, p. 139-150, Ed. by Gelfand. A. Hilger, London 1975.

[9] G. Harder, R.P. Langlands, M. Rapoport.- Algebraische Zyklen auf Hilbert-Blumenthal-Flächen. Jour. für die Reine und Angew. Math. 366 (1986), p. 53-120.

[10] F. Hirzebruch, D. Zagier.- Intersection Numbers of Curves on Hilbert Modular Surfaces. Inv. Math. 36 (1976), p. 57-113.

[11] Ch. Klingenberg.- Die Tate-Vermutungen für Hilbert-Blumenthal-Flächen. Dissertation, Bonn 1986.

[12] J. Labesse, R.P. Langlands.- L-Indistinguishability for SL(2). Can. J. 31, N° 4 (1979), p. 726-785.

[13] R.P. Langlands.- Base Change for GL(2). PUP Annals of Mathematics Studies N° 96.

[14] V.K. Murty, D. Ramakrishnan.- Period Relations and the Tate Conjecture for Hilbert Modular Surfaces. Preprint (Sept. 1986)

[15] T. Oda.- Periods of Hilbert Modular Surfaces. Birkhäuser, PM vol. 19, Boston 1982.

[16] D. Ramakrishnan.- Arithmetic of Hilbert-Blumenthal surfaces. Preprint (1986).

[17] D. Ramakrishnan.- Periods of Integrals Arising from K_1 of Hilbert-Blumenthal surfaces. Preprint (1985).

[18] M. Rapoport.- Compactification de l'espace de modules de Hilbert-Blumenthal. Comp. Math. 36 (1978), p. 255-335.

[19] G. Shimura.- On canonical models of arithmetic quotients of bounded symmetric domains. Ann. of Math. 91 (1970), p. 114-222.

[20] J. Tate.- Algebraic Cycles and Poles of Zetafunctions. Proc. Conf. Purdue Univ. 1963. Ed. by Shilling. Harper and Row, New York.

Christoph KLINGENBERG
Mathematisches Institut
der Universität zu Köln
WEYERTAL 86-90
D-5000 KOLN 41

SUR LA RAMIFICATION DES EXTENSIONS INFINIES DES CORPS LOCAUX

François LAUBIE

Introduction.

Nous nous proposons d'établir un bilan de la théorie de la ramification des extensions infinies des corps locaux principalement axé autour du problème suivant : soit L une extension algébrique infinie d'un corps local; définir et étudier les filtrations de ramification des extensions galoisiennes de L. En fait, dans cette situation, il y a deux théories de la ramification qui ne coïncident pas. L'une d'entre elles est dûe à J. Herbrand (1931-32) l'autre à Kawada, Satake et Tamagawa (1952); cette dernière a été renouvelée par J.-M. Fontaine et J.-P. Wintenberger grâce à leur théorie du corps des normes (1978-1981).

C'est elle que nous allons exposer en restreignant, le plus souvent, notre propos à l'étude des p-extensions totalement ramifiées.

Enfin, dans une dernière partie, nous présentons une vieille technique pour étudier la ramification, qui donne toujours des résultats. Il s'agit d'une distance ultramétrique sur certains ensembles d'extensions de corps, déja utilisée par Herbrand pour fonder sa théorie de la ramification et par Greenberg pour étudier les $\mathbb{Z}_p$-extensions des corps de nombres. Seule cette dernière partie contient des démonstrations complètes.

I - Théorie classique de la ramification (voir [10], ch. IV).

Soit K un corps local et soit L une extension galoisienne finie de K de groupe de Galois G. Pour tout $\sigma\in G$, on note $i_L(\sigma)=\inf(v(\sigma(x)-x)-1)$ où x parcourt l'anneau de valuation de L et où v désigne la valuation discrète de L telle que $v(L^*)=\mathbb{Z}$. En numérotation inférieure les groupes de ramification de G sont les $G_x=\{\sigma\in G \mid i_L(\sigma)\geq x\}$, $(x\in[-1,+\infty[)$. La fonction φ de Hasse-Herbrand est définie pour $x\geq 0$ par $\varphi_{L/K}(x)=\int_0^x \frac{dt}{(G_o:G_t)}$; la bijection réciproque se note $\psi_{L/K}$ et la numérotation supérieure des groupes de ramification est définie par $G^x=G_{\psi_{L/K}(x)}$ ou $G_x=G^{\varphi_{L/K}(x)}$ pour tout

$x\geq 0$; on a alors $\varphi_{L/K}(x)=\int_0^x (G^o:G^t)dt$. On note u_K la fonction d'ordre de la filtration $(G^x)_{x\geq 0}$: $\sigma\in G^x \Longleftrightarrow u_K(\sigma)\geq x$; on a $u_K=\varphi_{L/K}\circ i_L$.

Soit M une extension galoisienne de K contenue dans L et soit H le groupe de Galois de L/M. On a la formule de transitivité des applications de Hasse-Herbrand : $\varphi_{M/K}=\varphi_{L/K}\circ\varphi_{L/M}$ et le théorème de Herbrand : $(G/H)^x=G^xH/H$, pour tout $x\in\mathbb{R}^+$.

Le théorème de Herbrand permet de définir la filtration de ramification d'une extension galoisienne infinie d'un corps local en numérotation supérieure : supposons maintenant que l'extension L/K soit infinie, pour tout $x\geq 0$, on pose $G^x=\varprojlim \mathrm{Gal}(M/K)^x$ où M parcourt l'ensemble des extensions galoisiennes finies de K contenues dans L.

Cette filtration jouit notamment des propriétés suivantes :

- théorème de Herbrand : pour toute extension galoisienne M de K contenue dans L et pour tout $x\geq 0$ on a

$$\mathrm{Gal}(M/K)^x=G^x\mathrm{Gal}(L/M)/\mathrm{Gal}(L/M);$$

- continuité à gauche : pour tout $x\geq 0$, on a $G^x=\bigcap_{0\leq y<x} G^y$;

- théorème de Hasse-Arf : soit $\mathcal{U}_{L/K}=\{x\in\mathbb{R}^+ \mid \forall\epsilon>0,\ G^{x+\epsilon}\neq G^x\}$, si L/K est abélienne alors $\mathcal{U}_{L/K}\subset\mathbb{N}$.

Mais cette théorie appelle deux remarques :

1) Doit-on définir un groupe de ramification de L/K
   - comme un G^x pour $x\geq 0$?
   - ou comme un sous-groupe distingué fermé H de G tel que pour toute extension galoisienne finie M de K contenue dans L, H Gal(L/M)/Gal(L/M) soit un groupe de ramification de M/K ?

   Un tel sous-groupe H de G est de l'une des deux formes suivantes : $H=G^x$ pour un $x\geq 0$ ou $H=\overline{\bigcup_{y>x} G^y}$ pour un $x\geq 0$, où la barre désigne l'adhérence dans G, (Kawada, [3]).

2) De même, doit-on définir un saut supérieur de ramification de L/K
   - comme un $u\geq 0$ tel que $\forall\epsilon>0$, $G^{u+\epsilon}\neq G^u$ c'est-à-dire un élément de $\mathcal{U}_{L/K}$?

- ou comme un $u \geq 0$ tel qu'il existe un sous-groupe ouvert distingué H de G vérifiant $\forall \epsilon > 0$, $(G/H)^{u+\epsilon} \neq (G/H)^u$? On note $\mathcal{V}_{L/K}$ leur ensemble.

On a $\mathcal{V}_{L/K} \subset \mathbb{Q}$; soit $\mathcal{D}_{L/K}$ l'ensemble des limites des suites décroissantes d'éléments de $\mathcal{V}_{L/K}$, on a alors $\mathcal{U}_{L/K} = \mathcal{V}_{L/K} \cup \mathcal{D}_{L/K}$ (Kawada, [3]).

<u>Exemples</u> : a) si L est une clôture algébrique de K, alors $\mathcal{U}_{L/K} = \mathbb{R}^+$ et $\mathcal{V}_{L/K} = \mathbb{Q}^+$;

b) si L est une extension abélienne maximale de K alors $\mathcal{U}_{L/K} = \mathcal{V}_{L/K} = \mathbb{N}$;

c) si K est une extension finie de $\mathbb{Q}_p$ à corps résiduel $\mathbb{F}_q$ et si L est une extension abélienne maximale d'une extension abélienne maximale de K, alors

$$\mathcal{U}_{L/K} = \mathcal{V}_{L/K} = \{0, \frac{1}{q-1}, \frac{2}{q-1}, \ldots, 1 - \frac{1}{q-1}, 1, 1 + \frac{1}{q^2-q}, \ldots, 2 - \frac{1}{q^2-q}, 2, 2 + \frac{1}{q^3-q^2}, \ldots\}.$$

II - <u>Généralisations</u>.

Il est naturel de tenter de généraliser la théorie classique dans les deux directions suivantes :

- définir la ramification d'une extension galoisienne L/K où K est une extension infinie d'un corps local;

- définir la ramification d'un groupe infini d'automorphismes continus d'un corps local d'égales caractéristiques.

Il en résulte deux théories étroitement liées par la théorie des corps de normes de Fontaine et Wintenberger.

En ce qui concerne la première direction, il y a deux écoles :

1) La généralisation de J. Herbrand : il s'agit de deux articles ([2], [2']) parus en 1931 et 1932 après la mort de J. Herbrand et présentés par C. Chevalley et E. Noether. J. Herbrand construit ici une théorie de la ramification pour toutes les extensions L/K où K est une extension algébrique quelconque d'un corps de nombres. Malheureusement, comme Herbrand lui-même l'a remarqué, dans certaines situations intéressantes d'un point de vue arithmétique, cette théorie se trivialise. C'est pour cette raison qu'elle a été jusqu'à présent délaissée au profit de :

2) La généralisation de Kawada, Satake et Tamagawa : il s'agit de trois articles ([3], [8], [11]) parus en 1951 et 1952 et, aux dire de leurs auteurs, inspirés par A. Weil à la suite de son séjour à Tokyo. Contrairement à celle de Herbrand, cette théorie permet de construire une filtration de ramification pour les extensions galoisiennes L/K seulement lorsque K appartient à une certaine classe d'extensions de corps locaux; mais,

lorsqu'elle s'applique, elle fournit autant d'informations sur l'arithmétique de L/K que dans le cas où K est un corps local. Depuis 1978, des travaux de J.-M. Fontaine et J.-P. Wintenberger ont éclairé cette théorie d'un jour nouveau.

En ce qui concerne la deuxième direction de recherche, c'est-à-dire : définir la ramification des groupes infinis d'automorphismes des corps locaux, son histoire commence avec une conjecture d'A. Grothendieck résolue par S. Sen en 1969. Voici de quoi il s'agit : soit $K=\mathbb{F}_q((T))$ un corps local d'égales caractéristiques p (de sorte qu'il existe des automorphismes de K d'ordre infini) et soit σ un $\mathbb{F}_q$-automorphisme continu de K; on définit le nombre entier $i_K(\sigma)$ comme précédemment et on dit que σ est sauvagement ramifié si $i_K(\sigma)\geq 1$. Grothendieck a conjecturé et S. Sen a prouvé que, pour tout automorphisme sauvagement ramifié σ de K et pour tout entier $n\geq 1$, on a $i_K(\sigma^{p^{n-1}}) \equiv i_K(\sigma^{p^n})$ mod p^n; cela généralise le théorème de Hasse-Arf pour les extensions cycliques.

En fait, comme l'ont montré J.-M. Fontaine et J.-P. Wintenberger, cette généralisation du théorème de Hasse-Arf provient d'une correspondance fonctorielle entre certains groupes d'automorphismes de corps locaux qui sont des groupes de Lie p-adiques et certaines extensions galoisiennes de corps locaux à groupes de Galois des groupes de Lie p-adiques ([13], [14]).

L'idée maîtresse des théories de Kawada-Satake-Tamagawa et de Fontaine-Wintenberger, en ce qui concerne la ramification, est de déterminer les extensions de corps locaux dans lesquelles cohabitent les numérotations supérieures et inférieures de leurs groupes de ramification. Peut-être est-il utile d'exposer ce que l'on entend par là; pour des raisons de commodité d'exposition, nous restreignons dorénavant notre étude aux p-extensions totalement ramifiées donc, en particulier, aux pro-p-groupes.

III - <u>Numérotations supérieures et inférieures</u>.

Soit G un pro-p-groupe. On appelle filtration profinie de G tout ensemble $\mathcal{F}$ des sous-groupes ouverts distingués de G, totalement ordonné par l'inclusion et tel que $\bigcap_{F\in\mathcal{F}} F = (1)$.

Soit H un sous-groupe fermé de G. L'ensemble des $F\cap H$, F parcourant $\mathcal{F}$, est une filtration profinie de H dite induite par $\mathcal{F}$ sur H. Si, de plus, H est distingué dans G, l'ensemble des FH/H, F parcourant $\mathcal{F}$, est une flitration profinie de G/H dite filtration quotient de $\mathcal{F}$ sur G/H.

On appelle numérotation de la filtration profinie $\mathcal{F}$ de G, toute application ν de G-{1} dans $\mathbb{R}^+$, prolongée par $\nu(1)= \sup_{\sigma\neq 1} \nu(\sigma)$ telle que $\mathcal{F}$ coïncide avec la famille des parties infinies de G de la forme $\nu^{-1}([x,\nu(1)])$, x parcourant $[0,\nu(1)[$. Une numérotation ν de $\mathcal{F}$ est donc une fonction centrale sur G vérifiant : pour tous σ, $\tau\in G$ - $\nu(\sigma) \leq \nu(1)$

- $\nu(\sigma) = \nu(1) \Longrightarrow \sigma = 1$
- $\nu(\sigma\tau^{-1}) \geq \min(\nu(\sigma),\nu(\tau))$;

de plus, le groupe topologique G s'identifie alors à $\varprojlim G/\nu^{-1}([x,\nu(1)])$, x parcourant $[0,\nu(1)[$.

Etant donnée une numérotation ν de la filtration profinie $\mathcal{F}$ de G, on définit les fonctions φ et ψ de Hasse-Herbrand associée à ν par $\varphi_\nu(x)=\int_0^x \frac{dt}{(G:\nu^{-1}([t,\nu(1)]))}$ et $\psi_\nu(x)=\int_0^x (G:\nu^{-1}([t,\nu(1)]))dt$ pour tout $x\in[0,\nu(1)[$. Comme φ_ν et ψ_ν sont strictement croissantes, $\nu^+=\varphi_N\circ\nu$ et $\nu^-=\psi_\nu\circ\nu$ sont aussi des numérotations de $\mathcal{F}$ et on a $\nu^{+-}=\nu^{-+}=\nu$ ou, autrement dit, $\psi_{\nu^+}=\varphi_\nu^{-1}$ et $\varphi_{\nu^-}=\psi_\nu^{-1}$. Enfin, pour tout $x\in[0,\nu(1)]$, on a $\varphi_\nu(x)=\int \inf(\nu(\sigma),x)d_G\sigma$ où d_G désigne la mesure de Haar sur G.

Etant donnée une filtration profinie $\mathcal{F}$ de G, on appelle double numérotation arithmétique de $\mathcal{F}$ tout couple (i,u) de numérotations de $\mathcal{F}$ vérifiant $u=i^+$ (ou $i=u^-$) et $i(1)=u(1)=+\infty$. Pour se ramener à la terminologie de Fontaine et Wintenberger [1], on appelle filtration arithmétiquement profinie (en abrégé : APF) de G la donnée d'une filtration profinie $\mathcal{F}$ de G et d'une double numérotation arthmétique (i,u) de $\mathcal{F}$; les seules fonctions de Hasse-Herbrand associées à une filtration APF $(\mathcal{F},i,u)$ qu'il soit utile de considérer, sont φ_i et $\psi_u=\varphi_i^{-1}$; ce sont des bijections linéaires par morceaux de $\mathbb{R}^+$ sur $\mathbb{R}^+$.

Etant donnés une filtration APF $(\mathcal{F},i_G,u_G)$ de G et un sous-groupe fermé H de G, on note i_H la restriction de i_G à H, $\varphi_H=\varphi_{i_H}$, $u_H=i_H^+=\varphi_H\circ i_H$ et $\psi_H=\psi_{u_H}=\varphi_H^{-1}$; $(\mathcal{F}_H,i_H,u_H)$ est alors une filtration APF sur H appelée filtration APF induite par $(\mathcal{F},i_G,u_G)$ sur H. Pour tout $\sigma\in H$ on a $u_H(\sigma)=\int \inf(i_G(h),i_G(\sigma))d_Hh$ où d_H désigne la mesure de Haar sur H.

Supposons de plus que H soit distingué dans G. Pour tout $\sigma\in G$ d'image $\bar{\sigma}\in G/H$, on pose $u_{G/H}(\bar{\sigma})=\sup_{h\in H} u_G(\sigma h)$; on note également $\psi_{G/H}=\psi_{u_{G/H}}$, $i_{G/H}=u_{G/H}^-=\psi_{G/H}\circ u_{G/H}$ et $\varphi_{G/H}=\varphi_{i_{G/H}}=\psi_{G/H}^{-1}$; $(\mathcal{F}_{G/H},$

$i_{G/H}, u_{G/H})$ est une filtration APF sur G/H appelée filtration APF quotient de $(\mathcal{F}, i_G, u_G)$ sur G/H. On a alors $\varphi_{G/H} = \varphi_G \circ \varphi_H$ et, pour tout $\sigma \in G$, $i_{G/H}(\bar{\sigma}) = \int i_G(\sigma h) d_H h = \varphi_H(\sup_{h \in H} i_G(\sigma h))$.

Signalons que si I et J sont deux sous-groupes convenables de G alors $u_{J/I\cap J}$ (resp. $i_{J/I\cap J}$, $u_{(G/I)/(J/I)}$, $i_{(G/I)/(J/I)}$) s'identifie à $u_{IJ/I}$ (resp. $i_{IJ/I}$, $u_{G/J}$, $i_{G/J}$) par l'isomorphisme canonique de $J/I\cap J$ sur IJ/I (resp. de $(G/I)/(J/I)$ sur G/J).

Soit toujours $(\mathcal{F}, i_G, u_G)$ une filtration APF du pro-p-groupe G et soit (H_n) une suite décroissante de sous-groupes ouverts distingués de G telle que $\cap H_n = (1)$; pour tout $n \in \mathbb{N}$ et tout $\sigma \in G$, on note σ_n l'image de σ dans G/H_n. Pour tout n assez grand, on a $i_{G/H_n}(\sigma_n) = i_G(\sigma)$.

Soit F un sous-groupe fini distingué de G; il existe $n_0 \in \mathbb{N}$ tel que, pour tout $n \geq n_0$, on ait $F \cap H_n = (1)$ et il existe $n_1 \geq n_0$ tel que, pour tout $n \geq n_1$, on ait :

$$i_{FH_n/H_n}(\sigma_n) = i_G(\sigma).$$

Signalons, pour clore ce chapitre, qu'il n'est pas difficile de généraliser les filtration APF aux groupes profinis G qui admettent un p-sous-groupe ouvert distingué et même aux ensembles de classes à gauche (ou à droite) G/H où G est un groupe profini comme ci-dessus et H un sous-groupe fermé de G.

IV - <u>Extensions arithmétiquement profinies</u>.

On rappelle que l'on se cantonne ici à l'étude des p-extensions galoisiennes infinies totalement ramifiées des corps locaux de caractéristique résiduelle p; cela restera sous-entendu par la suite. La théorie générale des extensions APF se trouve dans [12].

On dit qu'une extension L/K de groupe de Galois G est APF si, pour tout $x \in \mathbb{R}^+$, G^x est d'indice fini dans G; on pose alors $u_K(\sigma) = \sup_{\sigma \in G^x} x$ pour tout $\sigma \in G$, et, avec les notations du paragraphe précédent, $i_L = u_K^-$. Dans ces conditions, $(\{G^x\}, i_L, u_K)$ est une filtration APF de G et, pour toute sous-extension galoisienne M/K de L/K, la filtration APF quotient de $(\{G^x\}, i_L, u_K)$ sur Gal(M/K) coïncide avec la filtration de ramification de M/K.

<u>Exemples</u> : a) les extensions abéliennes totalement ramifiées des extensions finies de $\mathbb{Q}_p$ sont APF;

b) les extensions galoisiennes totalement ramifiées de corps locaux dont le groupe de Galois est un groupe de Lie p-adique de dimension finie non nulle, sont APF (résultat dû à Sen [9], dans le cas d'inégales caractéristiques et à Wintenberger [14] dans le cas d'égales caractéristiques).

Lorsque L est une extension APF d'un corps local K, on définit la ramification des extensions galoisiennes de L de la manière suivante.

Soit tout d'abord M une extension galoisienne finie de L de groupe de Galois G. Il existe $\alpha \in M$ tel que $M = L(\alpha)$. Soit (K_n) une suite croissante d'extensions finies de K telle que $L = \cup K_n$; pour tout n assez grand, le corps $M_n = K_n(\alpha)$ est galoisien sur K_n et les extensions L et M_n de K_n sont linéairement disjointes de sorte qu'il existe un isomorphisme naturel φ_n de G sur $\mathrm{Gal}(M_n/K_n)$.

En fait, pour tout couple $n \geq m$ d'entiers assez grands, l'isomorphisme canonique $\varphi_{n,m} = \varphi_m \circ \varphi_n^{-1}$ est un isomorphisme de ramification, c'est-à-dire satisfait aux conditions :

$$\varphi_{n,m}(\mathrm{Gal}(M_n/K_n)^x) = \mathrm{Gal}(M_m/K_m)^x \quad \text{pour tout} \quad x \in \mathbb{R}^+.$$

On définit alors la filtration de ramification de G par $G^x = \varphi_n^{-1}(\mathrm{Gal}(M_n/K_n)^x)$ pour tout $x \in \mathbb{R}^+$, où n désigne un entier assez grand.

Il est clair que le théorème de Herbrand fonctionne pour ces filtrations. Donc, si maintenant M est une extension infinie de L, on peut définir la filtration de ramification de son groupe de Galois G en posant $G^x = \varprojlim \mathrm{Gal}(E/L)^x$ pour tout $x \in \mathbb{R}^+$, E/L parcourant les sous-extensions galoisiennes finies de M/L.

Cette théorie est très semblable à la théorie classique de la ramification; en particulier, les théorèmes de Herbrand, de Hasse-Arf ... s'appliquent sans changement.

Si L est une extension APF d'un corps local K, on dit encore qu'une extension galoisienne de L est APF si ses groupes de ramification sont d'indice finis dans son groupe de Galois. Dans [8], Satake signale qu'il a échoué dans sa tentative de répondre à la question suivante : une extension APF d'une extension APF est-elle une extension APF ? Dans [12], Wintenberger y répond affirmativement, plus précisément :

<u>Théorème</u> : Soit L une extension galoisienne APF d'un corps local K et soit M une extension galoisienne de K contenant L. Pour que M/K soit APF, il faut et il suffit que M/L le soit et dans ce cas la filtration de ramification de M/L est la filtration APF induite sur $\mathrm{Gal}(M/L)$ par la filtration de ramification de M/K.

La démonstration donnée par Wintenberger n'est pas directe : elle passe par l'étude du foncteur "corps de normes"; elle a cependant l'avantage d'expliquer la similitude entre cette théorie et la théorie classique de la ramification.

V - <u>Groupes d'automorphismes arithmétiquement profinis</u>.

Soit K un corps local d'égales caractéristiques p et soit $Aut_1(K)$ le groupe des automorphismes sauvagement ramifiés de K. Pour tout nombre réel $x \geq 1$, on note $Aut_x(K)$ le sous-groupe de $Aut_1(K)$ formé des σ tels que $i_K(\sigma) \geq x$. Cela définit une filtration, appelée filtration de ramification, pour laquelle $Aut_1(K)$ est un groupe séparé complet. Pour tout sous-groupe fermé G de $Aut_1(K)$ on note (G_x) la filtration induite; pour que G soit un sous-groupe compact de $Aut_1(K)$ il faut et il suffit que, pour tout $x>1$, G_x soit d'indice fini dans G et, lorsqu'il en est ainsi, G s'identifie au pro-p-groupe $\varprojlim G/G_n$, n parcourant $\mathbb{N}$.

Avec les notations du paragraphe III, on pose $i_G = i_K$ et $u_G = i_G^+$; pour que $(\{G_x\}, i_G, u_G)$ soit une filtration APF de G, il faut et il suffit que $\lim_{x \to +\infty} \varphi_G(x) = +\infty$ où $\varphi_G = \varphi_{i_G}$ désigne la fonction de Hasse-Herbrand.

Cette théorie présente aussi un grand nombre de similitudes avec la théorie classique de la ramification. Cependant, en ce qui concerne le théorème de Herbrand, la filtration APF quotient sur G/H où H désigne un sous-groupe distingué, fermé et infini de G, ne possède pas a priori d'interprétation arithmétique (le corps des invariants de H étant alors réduit au corps résiduel de K). La théorie du corps des normes de Fontaine et Wintenberger fournit, dans certains cas, une telle interprétation.

VI - <u>La théorie du corps des normes</u>.

Soit K une extension APF d'un corps local K. On note $X_K^*(L)$ le groupe multiplicatif $\varprojlim E^*$ où E parcourt l'ensemble des sous-extensions galoisiennes finies de L/K et où pour deux telles extensions $E' \supset E$, l'application de transition de E'^* à E^* est la norme $N_{E'/E}$; on pose également $X_K(L) = X_K^*(L) \cup \{0\}$ et on prolonge la multiplication sur $X_K^*(L)$ à $X_K(L)$ en considérant 0 comme un annulateur. Il existe alors une addition qui confère à $X_K(L)$ une structure de corps de caractéristique p, et une valuation discrète qui en fait un corps local dont le corps résiduel est isomorphe à celui de K.

Soit maintenant une autre extension APF L'/K' et soit τ un plongement continu fini séparable de L dans L' tel que K' soit une extension finie de $\tau(K)$. Il existe alors un plongement continu $W(\tau)$ de $X_K(L)$ dans $X_{K'}(L')$ qui permet d'identifier $X_{K'}(L')$ à une extension séparable de $X_K(L)$ de degré $[L':\tau(L)]$. Si $K'=K$ et si τ est un K-plongement de L dans L', $W(\tau)$ se note $X_K(\tau)$; ce qui permet de considérer X_K comme un foncteur de la catégorie dont les objets sont les extensions APF infinies de K et les flèches les K-plongements continus, finis et séparables dans la catégorie dont les objets sont les corps locaux de caractéristique p et les flèches les plongements continus finis et séparables. Le foncteur X_K est fidèle (pour plus de précisions voir [1], [12], [13] et [14]).

De plus, si G désigne le groupe de Galois de L/K alors $X_K(G)$ est un groupe d'automorphismes de $X_K(L)$ et le foncteur X_K respecte la ramification; autrement dit, pour tout $x\in\mathbb{R}^+$, on a $X_K(G_x)= (X_K(G))_x$.

Etant donnée une extension APF L/K, le foncteur X_K permet de construire une équivalence de catégories $X_{L/K}$ de la catégorie dont les objets sont les extensions algébriques séparables de L et les flèches les L-plongements continus sur la catégorie dont les objets sont les extensions algébriques séparables de $X_K(L)$ et les flèches les $X_K(L)$-plongements continus.

De plus le foncteur $X_{L/K}$ respecte la ramification autrement dit : pour toute extension galoisienne M de L et pour tout $x\in\mathbb{R}^+$, on a $X_{L/K}(\mathrm{Gal}(M/L)^x)= \mathrm{Gal}(X_{L/K}(M)/X_K(K))^x$. Et, pour finir, le foncteur $X_{X_K(L)}\circ X_{L/K}$ est isomorphe à X_K; le théorème du paragraphe IV en résulte.

Soit, maintenant, X un corps local d'égales caractéristiques, soit G un groupe APF d'automorphismes de X et soit H un sous-groupe distingué fermé de G. Dans le cas où il existe une extension APF L d'un corps local K telle que $X_K(L)=X$ et $X_K(\mathrm{Gal}(L/K))=H$, nous pouvons donner une signification arithmétique à la filtration APF quotient sur G/H de la filtration de ramification de G : G/H s'identifie alors à un groupe d'automorphismes APF de K et cette filtration n'est autre que sa filtration de ramification.

La question la plus importante posée par cette théorie est la suivante : quels sont les groupes APF d'automorphismes de corps locaux qui correspondent à des groupes de Galois d'extensions APF de corps locaux par le foncteur "corps de normes" ? J.-P. Wintenberger a montré dans [13] que les groupes abéliens topologiquement de type fini sont dans ce cas, ce qui lui a permis de généraliser le théorème de Hasse-Arf à ces groupes; dans cette direction, il existe également des résultats partiels concernant les groupes d'automorphismes qui sont des groupes de Lie p-adiques compacts.

VII - Une métrique liée à la ramification des extensions de corps locaux; applications.

Soit K un corps local de caractéristique résiduelle $p>0$ et soit $\mathcal{E}_K$ l'ensemble des extensions galoisiennes de K contenues dans une clôture séparable fixée de K. Pour tout $E\in\mathcal{E}_K$ et tout $u\in[-1,+\infty[$ on désigne par E^u le corps des invariants du groupe de ramification $\mathrm{Gal}(E/K)^u$. Pour tous E_1, $E_2\in\mathcal{E}_K$ on pose $v(E_1,E_2)=\sup_{E_1^u=E_2^u} u$ et $d(E_1,E_2)=p^{-v(E_1,E_2)}$; d est une distance ultramétrique sur $\mathcal{E}_K$ pour laquelle l'ensemble $\mathcal{E}_K^u$ des $E\in\mathcal{E}_K$ tels que $E^u=E$ est une partie discrète de $\mathcal{E}_K$. Pour tous $u,v\in[-1,+\infty[$ tels que $u\geq v$, on note ρ_{uv} l'application qui à $E\in\mathcal{E}_K^u$ associe $E^v\in\mathcal{E}_K^v$; ces applications font de l'ensemble des $\mathcal{E}_K^u$, $(u\in[-1,+\infty[)$ un système projectif d'espaces discrets.

Proposition 1 : Les espaces topologiques $\mathcal{E}_K$ et $\varprojlim \mathcal{E}_K^u$ sont homéomorphes; en particulier l'espace métrique $\mathcal{E}_K$ est complet.

Démonstration : L'application ρ qui à $E\in\mathcal{E}_K$ associe $(E^u)_{u\in[-1,+\infty[}$ est un plongement continu de $\mathcal{E}_K$ dans $\varprojlim \mathcal{E}_K^u$. Soit $(E_u)_{u\in[-1,+\infty[}$ une suite projective avec $E_u\in\mathcal{E}_K^u$ pour tout $u\in[-1,+\infty[$; si $v\geq u$ alors $E_v\supset E_u$; on note $E=\bigcup_{u\in[-1,+\infty[} E_u$, $G=\mathrm{Gal}(E/K)$ et $H_u=\mathrm{Gal}(E/E_u)$. Les H_u $(u\geq -1)$ forment une suite décroissante de sous-groupes fermés de G telle que $\bigcap_{u\geq -1} H_u=(1)$ et, pour tout $v\in[-1,+\infty[$, G^v est un sous-groupe fermé de G. En vertu du théorème de Herbrand ([10], ch. IV, § 3, lemme 5), on a donc $G^v=\bigcap_{u\in[-1,+\infty[} \mathrm{Gal}(E/E_u^v)$, c'est-à-dire : $E^v=\bigcup_{u\in[-1,+\infty[} E_u^v=\bigcup_{u\geq v} E_u^v=E_v^v=E_v$. Il en résulte que l'application qui à $(E_u)_{u\in[-1,+\infty[}\in\varprojlim \mathcal{E}_K^u$ associe $\bigcup_{u\in[-1,+\infty[} E_u\in\mathcal{E}_K$ est l'homéomorphisme réciproque de ρ. ■

<u>Remarques</u> :

1. L'ensemble $\mathcal{E}_K^{-1}$ est réduit à $\{K\}$; pour tout $u\in]-1,0]$, $\mathcal{E}_K^u=\mathcal{E}_K^o$ est l'ensemble des extensions non ramifiées de K; $\mathcal{E}_K^+ = \bigcap_{u>0} \mathcal{E}_K^u$ est l'ensemble des extensions galoisiennes modérément ramifiées de K; enfin, pour tous $u,v\in\mathbb{R}^+$ tels que $u<v$, on a $\mathcal{E}_K^u \subsetneq \mathcal{E}_K^v$ [7].

2. La distance introduite par J. Herbrand ([2], p. 712-713) pour établir la théorie de la ramification des extensions infinies induit sur $\mathcal{E}_K$ une topologie strictement moins fine que la notre.

<u>Proposition 2</u> : Soit $\mathcal{F}_K$ l'ensemble des extensions galoisiennes finies de K. Si K est un corps local d'inégales caractéristiques alors $\mathcal{F}_K$ est un ouvert discret de $\mathcal{E}_K$.

<u>Démonstration</u> : Pour tout $F\in\mathcal{F}_K$, posons $e'_F = \frac{p\, e_F}{p-1}$ où e_F désigne l'indice de ramification absolu de F. L'ensemble $B(F,e'_F)$ des $E\in\mathcal{E}_K$ tels que $v(E,F)\geq e'_F$ est un voisinage de F dans $\mathcal{E}_K$. Montrons que $B(F,e'_F)=\{F\}$, ce qui prouvera notre assertion. Posons $F'=F^{e'_F}$; pour tout $E\in B(F,e'_F)$ on a $E^{e'_F}=F'$. Supposons que $E\neq F'$, alors $Gal(E/F')$ est un p-groupe fini ou un pro-p-groupe parce que $e'_F>1$ et il existe une sous-extension galoisienne E'/F' de E/F' de degré p. On a $e'_F = \frac{p\, e_F}{p-1} \geq \frac{p\, e_{F'}}{p-1} = \frac{e_{E'}}{p-1}$; de plus, pour tout $u > \frac{e_{E'}}{p-1}$, on a $Gal(E'/F')^u=(1)$ (voir [10], ch. IV, § 2, exercice 3), d'où $E'=E'^{e'_F}\subset E^{e'_F}=F'$ ce qui est impossible. Il s'ensuit que pour tout $E\in B(F,e'_F)$, on a $E=E^{e'_F}=F^{e'_F}=F$. ■

<u>Remarque</u> : L'adhérence $\overline{\mathcal{F}}_K$ de $\mathcal{F}_K$ dans $\mathcal{E}_K$ est constituée des extensions galoisiennes de K arithmétiquement profinies au sens de Fontaine et Wintenberger ([1], [12]).

Soit G un groupe fini ou profini. On dit que $E\in\mathcal{E}_K$ est une G-extension de K si les groupes G et $Gal(E/K)$ sont isomorphes. On note $\mathcal{R}_K[G]$ l'ensemble des G-extensions totalement ramifiées de K et $\mathcal{R}_K(G)$ l'ensemble des G/H-extensions totalement ramifiées de K, H parcourant l'ensemble des sous-groupes distingués fermés de G.

<u>Proposition 3</u> : Supposons que K soit une extension finie de $\mathbb{Q}_p$ et que G soit un groupe de Lie p-adique de dimension finie non nulle; alors $\mathcal{R}_K(G)$ est une partie compacte de $\mathcal{E}_K$.

Lemme 1 : Soit G un groupe de Lie p-adique compact, soit $\mathcal{G}$ un groupe profini et soit $(\mathcal{G}_x)_{x\in\mathbb{R}^+}$ une famille de sous-groupes ouverts distingués de $\mathcal{G}$ telle que $\bigcap_{x\in\mathbb{R}^+} \mathcal{G}_x=(1)$ et $\mathcal{G}_x \supset \mathcal{G}_y$ si $x\leq y$.

Supposons que pour tout $x\in\mathbb{R}^+$, $\mathcal{G}/\mathcal{G}_x$ soit isomorphe à un groupe quotient de G. Alors $\mathcal{G}$ est isomorphe à un quotient de G par un sous-groupe distingué fermé.

Démonstration : Le groupe de Lie p-adique compact G ne possède au plus qu'un nombre fini de sous-groupes ouverts d'indice donné; il en résulte que, pour tout $x\in\mathbb{R}^+$, l'ensemble Φ_x des homomorphismes surjectifs de G sur $\mathcal{G}/\mathcal{G}_x$ est fini. Pour tous $x,y\in\mathbb{R}^+$ tels que $x\leq y$, on note c_{yx} la surjection canonique de $\mathcal{G}/\mathcal{G}_y$ sur $\mathcal{G}/\mathcal{G}_x$ et π_{yx} l'application qui, à $\phi\in\Phi_y$, associe $c_{yx}\circ \phi\in\Phi_x$. Les applications π_{yx} organisent les Φ_x en un système projectif d'ensembles finis non vides. Un élément ϕ de $\varprojlim \Phi_x$, qui est non vide, s'identifie à un homomorphisme surjectif de G sur $\varprojlim \mathcal{G}/\mathcal{G}_x=\mathcal{G}$ de noyau $\bigcap_{x\in\mathbb{R}^+} \phi^{-1}(\mathcal{G}_x)$. ■

Démonstration de la proposition 3 : Pour tout $u\in\mathbb{R}^+$, on pose $\mathcal{R}^u_K(G)= \mathcal{R}_K(G) \cap \mathcal{E}^u_K = \{L^u; L\in\mathcal{R}_K(G)\}$; pour $u\geq v$, ρ_{uv} applique $\mathcal{R}^u_K(G)$ dans $\mathcal{R}^v_K(G)$; l'ensemble des $\mathcal{R}^u_K(G)$ est donc un système projectif pour les applications ρ_{uv} et $\varprojlim \mathcal{R}^u_K(G)$ s'identifie à l'adhérence $\overline{\mathcal{R}_K(G)}$ de $\mathcal{R}_K(G)$ dans $\mathcal{E}_K$. Autrement dit, tout $L\in\overline{\mathcal{R}_K(G)}$ s'écrit $L= \bigcup_{u\in\mathbb{R}^+} L^u$ avec $L^u \in \mathcal{R}^u_K(G)$ pour tout $u\in\mathbb{R}^+$. Il résulte alors du lemme 1 que $L\in\mathcal{R}_K(G)$ et donc que $\mathcal{R}_K(G)$ est une partie fermée de $\mathcal{E}_K$ homéomorphe à la limite projective d'espaces discrets $\varprojlim \mathcal{R}^u_K(G)$.

Montrons que, pour tout $u\in\mathbb{R}^+$, $\mathcal{R}^u_K(G)$ est un ensemble fini. Il en résultera que $\mathcal{R}_K(G)$ est compact.

Considérons d'abord le cas où G est un pro-p-groupe et supposons qu'il existe $u\in\mathbb{R}^+$ tel que $\mathcal{R}^u_K(G)$ soit infini. Tout $L\in\mathcal{R}^u_K(G)$ est une extension de K de degré fini [9]. Comme il n'y a qu'un nombre fini d'extensions de K de degré fini fixé, pour tout $n\in\mathbb{N}$, il existe $L_n\in\mathcal{R}^u_K(G)$ tel que $[L_n:K]\geq p^n$ et, comme $Gal(L_n/K)$ est un p-groupe, L_n possède au moins une sous-extension galoisienne de degré p^n sur

K. Appelons $\mathcal{L}_n$ l'ensemble des $L\in\mathcal{R}_K^u(G)$ tels que $[L:K]=p^n$ et donnons-nous une application f_n qui, à tout $L\in\mathcal{L}_{n+1}$, associe une sous-extension de L appartenant à $\mathcal{L}_n$. Cela suffit pour faire des $\mathcal{L}_n$ un système projectif d'ensembles finis non vides. Un élément M de la limite projective, qui est non vide, s'identifie à une extension galoisienne, infinie, totalement ramifiée de K, dont tous les sauts de ramification supérieurs sont $\leq u$ (théorème de Herbrand, loc. cit.) et dont le groupe de Galois est une groupe de Lie p-adique (lemme 1). Or une telle extension de K ne peut pas exister [9], donc l'ensemble $\mathcal{R}_K^u(G)$ est fini.

Passons maintenant au cas général. Soit H un sous-groupe ouvert de G, d'indice h, qui est un pro-p-groupe et soit $\{K_1,\dots,K_n\}$ l'ensemble des extensions de K de degré $\leq h$. Pour tout $L\in\mathcal{R}_K(G)$, donnons-nous un homomorphisme surjectif continu φ_L de G sur Gal(L/K) et appelons L_H le corps des invariants de $\varphi_L(H)$. Pour tout $u\in\mathbb{R}^+$, l'ensemble $\mathcal{R}_K^u(G)$ s'écrit comme la réunion $\mathcal{M}_o \cup \mathcal{M}_1 \cup \dots \cup \mathcal{M}_n$ où $\mathcal{M}_o$ désigne l'ensemble fini des $L\in\mathcal{R}_K^u(G)$ tels que $[L:K]<h$ et où, pour $1\leq i\leq n$, $\mathcal{M}_i$ désigne l'ensemble des L^u, $L\in\mathcal{R}_K(G)$, tels que $L_H=K_i$. D'après le théorème de Herbrand (loc. cit.), pour tout $i=1,\dots,n$, il existe $u_i\in\mathbb{R}^+$ tel que $\mathcal{M}_i \subset \mathcal{R}_{K_i}^{u_i}(H)$; les ensembles $\mathcal{M}_i$ sont donc finis et $\mathcal{R}_K^u(G)$ aussi.

<u>Corollaire 1</u> : On suppose en plus des hypothèses de la proposition 3 que le groupe G est abélien; alors $\mathcal{R}_K[G]$ est une partie compacte de $\mathcal{E}_K$.

<u>Démonstration</u> : Il suffit de montrer que $\mathcal{R}_K[G]$ est fermé dans $\mathcal{R}_K(G)$. Le groupe G est isomorphe à $g\times\mathbb{Z}_p^r$ où g est un groupe fini et r un entier.

Considérons d'abord le cas où $G\simeq\mathbb{Z}_p^r$. Soit $E\in\mathcal{R}_K(G)$. On rappelle que e'_K désigne $\frac{p\,e_K}{p-1}$ où e_K est l'indice de ramification absolu de K. D'après la théorie du corps de classes local, pour tout $u\geq e'_K$, le groupe de ramification $\mathrm{Gal}(E/K)^{u+e_K}$ est formé des puissances p-ièmes des éléments de $\mathrm{Gal}(E/K)^u$ [6]. Comme les groupes de ramification de E/K forment un système fondamental de voisinages de 1 dans Gal(E/K) [6], la dimension de Gal(E/K) est égale au nombre minimum de générateurs de $Gal(E/K)/Gal(E/K)^{e'_K}$. Supposons que $E\in\overline{\mathcal{R}_K[G]}$; il existe alors $L\in\mathcal{R}_K[G]$ tel que $v(E,L)\geq e'_K$. Comme $E\in\mathcal{R}_K(G)$, il existe un homomorphisme surjectif continu de $Gal(L/K)$, sur $Gal(E/K)$; mais, d'après ce qui précède, $Gal(L/K)$ et $Gal(E/K)$ ont même dimension r; donc ϕ est un isomorphisme et $E\in\mathcal{R}_K[G]$.

Passons maintenant au cas général : soit H un sous-groupe ouvert de G isomorphe à $\mathbf{Z}_p^r$. On désigne par $\{K_1,\dots,K_n\}$ l'ensemble des $E\in\mathcal{E}_K$ tels que $Gal(E/K)\simeq G/H$. On a alors $\mathcal{R}_K[G]\subset \mathcal{M}_1\cup\dots\cup\mathcal{M}_n$ avec $\mathcal{M}_i=\mathcal{R}_{K_i}[H]\cap\mathcal{E}_K$ pour tout $i=1,\dots,n$; $\mathcal{E}_K\cap\mathcal{E}_{K_i}$ est fermé dans $\mathcal{E}_K$ et $\mathcal{R}_{K_i}[H]$ est fermé dans $\mathcal{E}_{K_i}$; donc $\mathcal{M}_i$ est fermé dans $\mathcal{E}_K$ et $\overline{\mathcal{R}_K[G]}\subset(\mathcal{M}_1\cup\dots\cup\mathcal{M}_n)\cap\mathcal{R}_K(G)$. Il en résulte que le groupe de Galois de tout $L\in\overline{\mathcal{R}_K[G]}$ sur K est isomorphe à G et donc que $L\in\mathcal{R}_K[G]$. ■

On rappelle qu'un groupe de Lie p-adique G est dit simple si 1°) $\dim G\geq 2$, 2°) aucun sous-groupe ouvert de G ne possède de sous-groupes distingués fermés non triviaux de codimension non nulle; exemple : $SL_2(\mathbf{Z}_p)$.

<u>Corollaire 2</u> : Sous les hypothèses de la proposition 3, si le groupe de Lie p-adique G est simple alors $\mathcal{R}_K[G]$ est une partie compacte de $\mathcal{E}_K$.

<u>Démonstration</u> : Comme G est simple, $\mathcal{R}_K[G]$ est l'ensemble des $L\in\mathcal{R}_K(G)$ de degré infini sur K. D'après la proposition 2, $\mathcal{R}_K[G]$ est alors une partie fermée de $\mathcal{R}_K(G)$. ■

<u>Définitions</u> : Deux extensions galoisiennes de corps locaux L_1/K_1 et L_2/K_2 sont dites identiquement ramifiées, ou dans la même classe de ramification s'il existe un isomorphisme f entre leur groupes de Galois qui respecte leurs filtrations de ramification, c'est-à-dire tel que $f(Gal(L_1/K_1)^u)=Gal(L_2/K_2)^u$ pour tout $u\in[-1,+\infty[$; un tel f s'appelle un isomorphisme de ramification.

Le résultat suivant précise les théorèmes de M.A. Marshall concernant la ramification des extensions abéliennes des corps locaux [6].

<u>Théorème 1</u> : Soit A une groupe de la forme $\Gamma\times\mathbf{Z}_p^r$ où Γ est un groupe abélien fini, soit $e\in\mathbb{N}^*$ et soit $\mathcal{A}$ l'ensemble des A-extensions totalement ramifiées L/k, k parcourant un ensemble

quelconque de corps locaux d'inégales caractéristiques et d'indice de ramification absolu e. Alors la partition de $\mathcal{A}$ en classes de ramification est finie.

<u>Démonstration</u> : Montrons d'abord que, si K est une extension finie de $\mathbb{Q}_p$ d'indice de ramification e, alors la partition de $\mathcal{R}_k[A]$ en classes de ramification est finie. Pour tout $E\in\mathcal{E}_K$ et tout $u\in[-1,+\infty[$, on note $B(E,u)$ l'ensemble des $E'\in\mathcal{E}_K$ tels que $v(E,E')\geq u$; c'est la boule "fermée" centrée en E et de rayon p^{-u}. Soient $e'=\frac{e\,p}{p-1}$ et $e''=e'+e$; on a déjà vu que, pour tout $L\in\mathcal{R}_K[A]$ et tout $u\geq e'$, le groupe de ramification $Gal(L/K)^{u+e}$ est formé des puissances p-ième des éléments de $Gal(L/K)^u$.

Pour tout $L\in\mathcal{R}_K[A]$, on se donne un isomorphisme ϕ_L de $Gal(L/K)$ sur A et, pour tout $u\in\mathbb{R}^+$, on note $\phi_L(G(L/K)^u)=A_L^u$. La filtration $(A_L^u)_{u\in\mathbb{R}^+}$ est complètement déterminée par $(A_L^u)_{0\leq i\leq e''}$; plus précisément, si $u_1,\ldots,u_s$ désignent les sauts de cette filtration qui sont dans $[e',e''[$ alors l'ensemble des sauts $\geq e''$ est $\{u_t+me;\ t\in\{1,\ldots,s\},\ m\in\mathbb{N}^*\}$ et $A_L^{u_t+me}$ est formé des puissances p^m-ièmes des éléments de $A_L^{u_t}$.

Maintenant, comme $\mathcal{R}_K[A]$ est compact, il existe $L_1,\ldots,L_n\in\mathcal{R}_K[A]$ tels que $\mathcal{R}_K[A]\subset B(L_1,e'')\cup\ldots\cup B(L_n,e'')$ et il suffit de voir que, pour tout $i=1,\ldots,n$, la partition de $\mathcal{R}_K[A]\cap B(L_i,e'')$ en classes de ramification est finie. Il existe $\eta_i\in\mathbb{N}^*$ tel que, pour tout $L\in\mathcal{R}_K[A]\cap B(L_i,e'')$, on ait $(A:A_L^{e''})=\eta_i$; mais il n'y a qu'un nombre fini de sous-groupes ouverts distingués de A d'indice η_i et si $A_L^{e''}$ est l'un d'eux il n'y a qu'un nombre fini de filtrations partielles $(A_L^u)_{0\leq u<e''}$ telles que $A_L^{e''}\subset A_L^u$ pour tout $u\in[0,e'']$. Il en résulte qu'il n'y a qu'un nombre fini de filtrations de A susceptibles d'être isomorphes à la filtration de ramification du groupe de Galois d'un $L\in\mathcal{R}_K[A]\cap B(L_i,e'')$. Ainsi, dans le cas où $[K:\mathbb{Q}_p]<+\infty$, la partition en classes de ramification de $\mathcal{R}_K[A]$ est finie.

Maintenant si k est un corps local d'indice de ramification absolu e, on sait (voir [4] pour $p\neq 2$, et [5] pour $p=2$) que pour tout $L\in\mathcal{R}_k[A]$ il existe un corps local K à corps résiduel fini et d'indice de ramification absolu e tel que $\mathcal{R}_K[A]$ contienne une extension identiquement ramifiée à L/k et ceci achève la preuve. ■

Dans le cas des extensions galoisiennes totalement ramifiées à groupe de Galois un groupe de Lie p-adique non abélien, on obtient évidemment des résultats plus faibles que dans le cas abélien.

Définitions : Etant données deux extensions L_1 et L_2 dans $\mathcal{E}_K$, on dit qu'un isomorphisme de $Gal(L_1/K)$ sur $Gal(L_2/K)$ est cohérent s'il induit l'identité sur $Gal(L_1 \cap L_2/K)$. On dit qu'une partie $\mathcal{P}$ et $\mathcal{E}_K$ est cohérente si, pour tous L_1, $L_2 \in \mathcal{P}$, il existe un isomorphisme cohérent de $Gal(L_1/K)$ sur $Gal(L_2/K)$.

Remarque : On pourrait, si l'on voulait, affaiblir la définition d'un isomorphisme cohérent de $Gal(L_1/K)$ sur $Gal(L_2/K)$ en exigeant seulement qu'il induise sur $Gal(L_1 \cap L_2/K)$ un automorphisme de ramification.

Exemples 1. - L'ensemble des $\mathbb{Z}_p$-extensions de K est une partie cohérente de $\mathcal{E}_K$.

2. - Il est facile de construire deux extensions E_1 et E_2 de K de groupes de Galois isomorphes à $\mathbb{Z}/p^2 \times \mathbb{Z}/p$ telles que la partie $\{E_1, E_2\}$ de $\mathcal{E}_K$ ne soit pas cohérente.

Lemme 2 : Supposons que K soit une extension finie de $\mathbb{Q}_p$ et que G soit une groupe de Lie p-adique simple. Alors l'adhérence dans $\mathcal{E}_K$ d'une partie cohérente de $\mathcal{R}_K[G]$ est encore une partie cohérente de $\mathcal{R}_K[G]$.

Démonstration : Soit $\mathcal{P}$ une partie cohérente de $\mathcal{R}_K[G]$. Pour tout $v \in \mathbb{R}^+$, on pose $\mathcal{P}^v = \{L^v ; L \in \mathcal{P}\}$; pour $u \geq v$, ρ_{uv} applique $\mathcal{P}^u$ dans $\mathcal{P}^v$; l'ensemble des $\mathcal{P}^u$ est donc un système projectif pour les applications ρ_{uv} et $\varprojlim \mathcal{P}^u$ s'identifie à l'adhérence $\overline{\mathcal{P}}$ de $\mathcal{P}$ dans $\mathcal{E}_K$. Autrement dit, tout $L \in \overline{\mathcal{P}}$ s'écrit $L = \bigcup_{u \in \mathbb{R}^+} L_u$ avec $L_u \in \mathcal{P}^u$ pour tout $u \in \mathbb{R}^+$, c'est-à-dire un isomorphisme cohérent $\phi_{v,u}$ de $Gal(\tilde{L}_v/K)$ sur $Gal(\tilde{L}_u/K)$, donc il existe un homomorphisme surjectif $\overline{\phi}_{v,u}$ de $Gal(\tilde{L}_v/K)$ sur $Gal(L_u/K)$ qui induit l'identité sur $Gal(L_v/K)$. Comme les groupes $Gal(L_u/K)$ sont finis [9], par un procédé en tous points semblable à celui du lemme 1, on en déduit un homomorphisme surjectif continu $\phi_{L,v}$ de $Gal(\tilde{L}_v/K)$ sur $Gal(L/K)$ qui induit l'identité sur $Gal(L_v/K)$; mais comme $Gal(\tilde{L}_v/K)$ est un groupe de Lie p-adique simple, $\phi_{L,v}$ est un isomorphisme cohérent.

Soit maintenant M un autre élément de $\mathcal{P}$.

On a : $L \cap M = \left[\bigcup_{u\in\mathbb{R}^+} L_u\right] \cap \left[\bigcup_{u\in\mathbb{R}^+} M_u\right] = \bigcup_{u\in\mathbb{R}^+} (L_u \cap M_u) = L_{u_o} \cap M_{u_o}$ pour un certain $u_o \in \mathbb{R}^+$ parce que $L \cap M$ est une extension finie de K. Soit φ un isomorphisme cohérent de $Gal(\tilde{L}_{u_o}/K)$ sur $Gal(\tilde{M}_{u_o}/K)$; alors $\phi^{-1}_{L,u_o} \circ \varphi \circ \phi_{M,u_o}$ est un isomorphisme cohérent de $Gal(L/K)$ sur $Gal(M/K)$. ■

<u>Théorème 2</u> : Soit K une extension finie de Q_p et soit G une groupe de Lie p-adique simple. La partition en classes de ramification de toute partie cohérente de $\mathcal{R}_K[G]$ est finie.

<u>Démonstration</u> : Pour tout $E \in \mathcal{E}_K$ et tout $u \in \mathbb{R}^+$, on note $B(E,u)$ l'ensemble des $E' \in \mathcal{E}_K$ tels que $v(E,E') \geq u$. Pour tout $L \in \mathcal{R}_K[G]$, il existe $u_L \in \mathbb{R}^+$ vérifiant la propriété suivante : pour tout $L' \in \mathcal{R}_K[G] \cap B(L,u_L)$, tout isomorphisme cohérent de $Gal(L/K)$ sur $Gal(L'/K)$ est un isomorphisme de ramification ([5], prop. 4). Soit $\mathcal{P}$ une partie cohérente de $\mathcal{R}_K[G]$. Comme $\bar{\mathcal{P}}$ est compact, il existe $L_1,\ldots,L_n \in \bar{\mathcal{P}}$ tels que $\bar{\mathcal{P}} \subset B(L_1,u_{L_1}) \cup \ldots \cup B(L_n,u_{L_n})$. Comme l'ensemble $\bar{\mathcal{P}}$ est cohérent, toute extension L/K dans $\mathcal{P}$ est identiquement ramifiée à l'une des extensions $L_1/K,\ldots,L_n/K$. ■

Supposons que K soit un corps local d'inégales caractéristiques d'indice de ramification absolu e et que G soit un groupe de Lie p-adique. Shankar Sen a montré dans [9] que pour tout $L \in \mathcal{R}_K[G]$ il existe $v_L \in \mathbb{R}^+$ vérifiant la propriété suivante : pour tout $u \geq v_L$, le groupe de ramification $Gal(L/K)^{u+e}$ est formé des puissances p-ièmes des éléments de $Gal(L/K)^u$. Il serait intéressant de savoir si $\sup_{L\in\mathcal{R}_K[G]} v_L$ est nécessairement infini. Quoiqu'il ne réponde pas à la question, le résultat suivant apporte néanmoins une précision dans cette direction.

On rappelle que pour tout $u \in \mathbb{R}^+$, $\mathcal{R}^u_K(G)$ désigne l'ensemble des extensions galoisiennes finies totalement ramifiées de K dont le groupe de Galois est isomorphe à un groupe de quotient de G et dont le plus grand saut de ramification supérieur est $\leq u$.

<u>Proposition 5</u> : Soit K une extension finie de $\mathbb{Q}_p$ d'indice de ramification e et soit G un groupe de Lie p-adique. Il existe une

application w de $\mathbb{R}^+$ dans $\mathbb{R}^+$ vérifiant la propriété suivante : pour tout $x\in\mathbb{R}^+$, pour tout $F\in\mathcal{R}_K^{w(x)}(G)$ et pour tout $u\geq w(x)-x$, le groupe de ramification $Gal(F/K)^{u+e}$ est formé des puissances p-ièmes des éléments de $Gal(F/K)^u$.

<u>Démonstration</u> : Pour tout $L\in\mathcal{R}_K(G)$ on désigne par v_L la constante de Sen de l'extension de Lie L/K (voir plus haut) et, pour tout $u\in\mathbb{R}^+$, on note $B(L,u)$ la boule "fermée" de rayon p^{-u} et de centre L dans $\mathcal{E}_K$.

Comme $\mathcal{R}_K(G)$ est compact, pour tout $x\in\mathbb{R}^+$, il existe $L_1,\ldots,L_m\in\mathcal{R}_K(G)$ tels que $\mathcal{R}_K(G)\subset B(L_1,v_{L_1}+x+e)\cup\ldots\cup B(L_m,v_{L_m}+x+e)$. Soit $w(x)=w=\max_{1\leq i\leq m}(v_{L_i}+x+e)$; comme tous les $B(L_i,v_{L_i}+x+e)\cap\mathcal{R}_K(G)$ sont compacts, il existe $M_1,\ldots,M_n\in\mathcal{R}_K(G)$ tels que $\mathcal{R}_K(G)\subset B(M_1,w)\cup\ldots\cup B(M_n,w)$. Posons $\Gamma_i=Gal(M_i/K)$ pour tout $i=1,\ldots,m$ et, pour tout groupe H, notons $P(H)$ l'ensemble des puissances p-ièmes des éléments de H. Pour tout $u\in[w-x,w]$ on a $\Gamma_i^{u+e}=P(\Gamma_i^u)$; par suite, pour tout $F\in\mathcal{R}_K^w(G)$, il existe $i\in\{1,\ldots,n\}$ tel que $F\in B(M_i,w)$ et on a

$$Gal(F/K)^{u+e}=\Gamma_i^{u+e}/\Gamma_i^w=P(\Gamma_i^u/\Gamma_i^w)=P(Gal(F/K)^u).$$

L'application w répond donc à la question. ∎

Terminons cet article par une remarque concernant l'approximation des corps locaux d'égale caractéristique par ceux d'inégales caractéristiques.

Tout d'abord si L est une extension séparable infinie d'un corps local K, on note encore $\mathcal{E}_L$ l'ensemble des extensions galoisiennes de L. Il existe plusieurs théories de la ramification pour les éléments de $\mathcal{E}_L$ ([3], [8], [11]) mais elles coïncident dans les cas où L/K est une extension arithmétiquement profinie [12] (par exemple si L/K est une $\mathbb{Z}_p$-extension totalement ramifiée). On définit alors une distance ultramétrique sur $\mathcal{E}_L$ comme dans le cas où L est un corps local.

D'autre part, à tout corps local d'égale caractéristique X, on sait associer, par la théorie des corps de normes de Fontaine et Wintenberger, une $\mathbb{Z}_p$-extension totalement ramifiée L d'un corps local d'inégales caractéristiques K de sorte qu'il existe une correspondance fonctorielle $X_{L/K}$ entre les extensions séparables de L et celles de X ([12], [13]). Il ressort immédiatement de ces constructions que $X_{L/K}$ induit une isométrie de $\mathcal{E}_L$ sur $\mathcal{E}_X$.

Soit maintenant $F \in \mathcal{E}_L$ de degré fini sur L; il existe alors une extension finie K_n de K contenue dans L et $F_n \in \mathcal{E}_{K_n}$ tels que $F = LF_n$ et que F_n/K_n soit identiquement ramifiée à F/L (voir, par exemple [2], théorème 4). En conclusion toute extension galoisienne finie d'un corps local d'égale caractéristique est identiquement ramifiée à une extension galoisienne finie d'un corps local d'inégale caractéristique. Notons cependant que ce résultat peut être obtenu de façon plus directe et que sa réciproque est inexacte.

BIBLIOGRAPHIE

[1] J.-M. Fontaine, J.-P. Wintenberger.- Le "corps des normes" de certaines extensions algébriques de corps locaux, C.R. Acad. Sc. Paris, Série A, 288 (1979), 367-370.

[2] J. Herbrand.- Théorie arithmétiques des corps de nombres de degré infini I, Math. Ann., 106 (1932), 473-501.

[2'] J. Herbrand.- Théorie arithmétique des corps de nombres de degré infini II, Math. Ann., 108 (1933), 699-717.

[3] Y. Kawada.- On the ramification theory of infinite algebraic extensions, Ann. of Math., 58 (1953), 24-47.

[4] F. Laubie.- Groupes de ramification et corps résiduels, Bull. Sc. Math. 2-ième série, 105 (1981), 309-320.

[5] F. Laubie.- Sur la ramification des extensions de Lie, Compositio Math. 55 (1985), 253-262.

[6] M.A. Marshall.- Ramification groups of Abelian local field extensions, Can. J. Math. 23, N°2 (1971), 184-203.

[7] E. Maus.- On the jumps in the series of ramifications groups, Colloque de Théorie des Nombres, Bordeaux, 1969, Bull. Soc. Math. France, Mémoire 25 (1971), 127-133.

[8] I. Satake.- On a generalization of Hilbert theory of ramification, Sc. Paper Coll. of General Education Tokyo Univ. 2, N° 1 (1952), 25-39.

[9] S. Sen.- Ramification in p-adic Lie extensions, Invent. Math. 17 (1972), 44-50.

[10] J.-P. Serre.- Corps locaux, 2-ième ed., Hermann, Paris, 1968.

[11] T. Tamagawa.- On the theory of ramification groups and conductors, Jap. J. Math. 21 (1951), 197-215.

[12] J.-P. Wintenberger.- Le corps des normes de certaines extensions infinies de corps locaux; applications, Ann. Scient. Ec. Norm. Sup., 4-ième série, 16 (1983), 59-89.

[13] J.-P. Wintenberger.- Extensions abéliennes et groupes d'automorphismes de corps locaux, C.R. Acad. Sc. Paris, série A, 290 (1980), 201-204.

[14] J.-P. Wintenberger.- Extensions de Lie et groupes d'automorphismes des corps locaux de caractéristique p, C.R. Acad. Sc. Paris, série A, 288 (1979), 477-479.

François LAUBIE
UFR des Sciences de Limoges
Département de Mathématiques
123, avenue Albert Thomas
87060 LIMOGES Cédex

Une nouvelle démonstration du théorème d'isogénie, d'après D.V. et G.V. Choodnovsky

Michel LAURENT

1 - Introduction.

Le théorème d'isogénie dont il est question, s'énonce ainsi :

Théorème 1. Soient E et E' deux courbes elliptiques définies sur $\mathbb{Q}$ et ℓ un nombre premier. Les courbes elliptiques E et E' sont isogènes sur $\mathbb{Q}$, si et seulement si les modules galoisiens associés $V_\ell(E)$ et $V_\ell(E')$ sont isomorphes.

Rappelons que l'on désigne classiquement par $V_\ell(E)$ le $\mathbb{Q}_\ell$-espace vectoriel

$$V_\ell(E) = (\varprojlim_n E_{\ell^n}) \otimes_{\mathbb{Z}_\ell} \mathbb{Q}_\ell$$

et que l'action du groupe de Galois $G_{\mathbb{Q}}$ de $\overline{\mathbb{Q}}$ sur $\mathbb{Q}$ sur le groupe E_{ℓ^n} des points de ℓ^n-torsion de E, induit sur $V_\ell(E)$ une structure de $\mathbb{Q}_\ell[G_{\mathbb{Q}}]$-module. Pour plus de précisions sur ces définitions, on pourra consulter le livre de J.-P. Serre [14], dont nous suivrons les notations.

Le théorème ci-dessus est, bien sûr, contenu dans la conjecture de Tate démontrée par Faltings (cf. [6] et [15]) :

Théorème 2. Soient A et A' deux variétés abéliennes définies sur un corps de nombres K et ℓ un nombre premier. Désignons par $\mathrm{Hom}_K(A,A')$ le groupe des homomorphismes, définis sur K, de A dans A' et G_K le groupe de Galois de $\overline{K}$ sur K. La restriction aux points de ℓ^∞-torsion des morphismes de $\mathrm{Hom}_K(A,A')$, induit un isomorphisme canonique

$$\mathrm{Hom}_K(A,A') \otimes_{\mathbb{Z}} \mathbb{Q}_\ell \xrightarrow{\sim} \mathrm{Hom}_{G_K}(V_\ell(A), V_\ell(A')).$$

Nous nous proposons de donner ici une démonstration détaillée du théorème d'isogénie, suivant une note de D.V. et G.V. Choodnovsky [3]. L'ingrédient essentiel est un critère de dépendance algébrique qui peut être vu comme une généralisation du classique théorème de Schneider-Lang (cf. chap. 4 de [10]). La première tentative pour démontrer le théorème d'isogénie grâce à des arguments de transcendance est d'ailleurs due à S. Lang [11] et a été poursuivie par D. Masser [12] et D. Bertrand [1]. Le principe utilisé est assez simple : si E et E' désignent des courbes elliptiques non isogènes sur $\mathbf{C}$, associées à des fonctions de Weierstrass $\wp_1$ et $\wp_2$ algébriquement indépendantes, on peut espérer montrer, par une construction de fonction auxiliaire en $\wp_1$ et $\wp_2$, que les valeurs de ces fonctions en des points de division, engendrent des corps de nombres "largement" disjoints. Il s'ensuit en particulier que les modules $V_\ell(E)$ et $V_\ell(E')$ ne peuvent alors être isomorphes. D.V. et G.V. Choodnovsky procèdent de façon quelque peu différente. Ils ne travaillent pas directement avec des points de division, mais utilisent certains résultats de Honda sur la classification des groupes formels sur $\mathbf{Z}$, qu'ils interprètent analytiquement en termes de fonctions elliptiques.

2 - <u>Enoncés équivalents au théorème d'isogénie</u>.

Il me semble plus clair de présenter la preuve du théorème d'isogénie sous forme d'une suite d'équivalence.

<u>Théorème 3</u>. Soient E et E' deux courbes elliptiques définies sur $\mathbf{Q}$. Les propriétés suivantes sont équivalentes :

i) Il existe un nombre premier ℓ tel que les $G_{\mathbf{Q}}$-modules $V_\ell(E)$ et $V_\ell(E')$ sont isomorphes.

ii) Les fonctions L associées à E et à E' sont égales.

iii) Les groupes formels sur $\mathbf{Z}$ associés à E et à E' sont isomorphes.

iv) Les courbes E et E' sont isogènes sur $\mathbf{Q}$.

Précisons tout d'abord la signification de l'assertion iii). Désignons par $E_{\mathbf{Z}}$ le modèle de Néron sur $\mathbf{Z}$ de la courbe elliptique E. Le groupe formel $\hat{E}$ associé à E est par définition le complété formel de $E_{\mathbf{Z}}$ le long de l'origine de $E_{\mathbf{Z}}$. De façon plus concrète, choisissons un modèle de Weierstrass minimal global de E :

$$y^2 + a_1xy + a_3y = x^3 + a_2x^2 + a_4x + a_6, \quad a_i \in \mathbf{Z},$$

ainsi qu'un paramètre local t le long de l'origine de $E_{\mathbf{Z}}$, de telle sorte que x et y s'écrivent comme des séries formelles en t, à coefficients entiers;

$$x = t^{-2} + \dots , \qquad y = \pm t^{-3} + \dots$$

(on choisit habituellement le paramètre $t = -x/y$).

L'addition sur la courbe elliptique E s'exprime alors, en fonction du paramètre t, au moyen d'une loi de groupe formel $F \in \mathbf{Z}[[u,v]]$. On identifiera $\hat{E}$ à la classe d'isomorphisme sur $\mathbf{Z}$ des divers groupes formels F (tous analytiquement isomorphes) associés au choix d'un tel paramètre t. L'assertion iii) signifie alors que les classes $\hat{E}$ et $\hat{E}'$ associées aux courbes elliptiques E et E' sont égales. De plus, nous dirons qu'un groupe formel F appartenant à la classe $\hat{E}$ est <u>algébrique</u> s'il provient comme ci-dessus d'un paramètre local t qui est une fonction rationnelle sur E.

Rappelons que l'exponentielle $\exp_F$ d'une loi de groupe formel $F \in \mathbf{Z}[[u,v]]$ est l'unique série formelle de $\mathbf{Q}[[u]]$ caractérisée par les conditions :

$$\exp_F(u) = u + \dots , \qquad \exp_F(u+v) = F(\exp_F(u), \exp_F(v)).$$

Les séries exponentielles des lois de groupe formel algébriques de la classe $\hat{E}$, vérifient alors la propriété fondamentale suivante.

<u>Lemme</u>. Soit Λ le réseau des périodes de la forme différentielle $dx/(2y+a_1x+a_3)$. La série formelle $\exp_F$ est la série de Taylor à l'origine d'une fonction elliptique de réseau Λ.

En particulier, une telle série formelle $\exp_F$ admet un prolongement méromorphe dans tout le plan complexe.

<u>Preuve</u> : Désignons par ω l'une des deux formes différentielles $\pm dx/(2y+a_1x+a_3)$, choisie de telle sorte que ω s'écrive sous la forme

$$\omega = (1+\dots)dt$$

en fonction du paramètre local t. Dans l'isomorphisme analytique $E_{\mathbf{Z}}(\mathbf{C}) \xrightarrow{\sim} \mathbf{C}/\Lambda$, obtenu en intégrant ω, le paramètre local t s'identifie à une fonction elliptique de réseau Λ, régulière au voisinage de l'origine et de la forme $t(z)= z+\dots$. La série formelle $\exp_F$ est alors le développement de Taylor à l'origine de la fonction elliptique t. En effet, si u et v désignent des nombres complexes suffisamment proches de l'origine, on a par définition de la loi de groupe formel F :

$$t(u+v)= t(P+Q)= F(t(P),t(Q))= F(t(u),t(v)),$$

où P et Q désignent les points de $E_{\mathbf{Z}}(\mathbf{C})$ correspondant respectivement à $u \bmod \Lambda$ et à $v \bmod \Lambda$.

3 - <u>Preuve des équivalences i), ii), iii)</u>.

L'équivalence entre les propriétés i) et ii) est démontrée dans la proposition de la page IV.15 de [14] et résulte essentiellement de la semi-simplicité des $\mathbf{Q}_\ell[G_{\mathbf{Q}}]$-modules $V_\ell(E)$ et $V_\ell(E')$. le lien entre le module galoisien $V_\ell(E)$ et la fonction L de la courbe elliptique E, est le suivant. Si p désigne un nombre premier $\neq \ell$ et si la courbe elliptique E a bonne réduction en p, la représentation $G_{\mathbf{Q}} \longrightarrow GL(V_\ell(E))$ est non ramifiée en p. Le polynôme caractéristique X^2-a_pX+p de l'action sur $V_\ell(E)$ de n'importe quel élément σ de $G_{\mathbf{Q}}$, induisant un Frobenius au dessus de p, détermine alors le facteur local

$$L_p(s)= (1-a_pp^{-s}+p^{1-2s})^{-1}$$

de la fonction L associée à E. Nous indiquerons dans le paragraphe 6 des versions effectives de l'équivalence entre i) et ii).

Si F est une loi de groupe formel sur $\mathbf{Z}$, désignons par $\log_F$ la série formelle de $\mathbf{Q}[[u]]$, obtenue en inversant formellement la série exponentielle $\exp_F$ associée à F. L'implication de ii) vers iii) se déduit alors du résultat suivant, lié aux congruences d'Atkin et Swinnerton-Dyer, dû indépendamment à Cartier [2], Hill [8] et Honda [9].

<u>Théorème 4</u>. Soit E une courbe elliptique définie sur $\mathbf{Q}$ et $L(s)= \Sigma\, a_n n^{-s}$, sa fonction L. Il existe un groupe formel F dans la classe $\hat{E}$ tel que

$$\log_F(u)= \Sigma\, a_n u^n/n.$$

Remarquons que cette loi de groupe F n'a aucune raison d'être algébrique relativement à la courbe elliptique E.

Il s'ensuit en particulier que si les fonctions L des courbes E et E' sont égales, il existe deux lois de groupe formel sur $\mathbf{Z}$, F et F' appartenant respectivement à $\hat{E}$ et à $\hat{E}'$, telles que

$$\log_F= \log_{F'},$$

et donc $$F = F'.$$

L'implication inverse se démontre par réduction modulo p. Soit F une loi de groupe formel appartenant à la classe $\hat{E}= \hat{E}'$ et soit p un nombre premier tel que les courbes elliptiques E et E' aient bonne réduction en p. Désignons par F_p la loi de groupe formel sur $\mathbf{Z}/p\mathbf{Z}$, obtenue par réduction modulo p des coefficients de F. La hauteur h de F_p est égale à 1 ou 2, et la $\mathbf{Z}_p$-algèbre engendrée dans l'anneau des endomorphismes de F_p par l'endomorphisme de Frobenius π, est un ordre dans une extension totalement ramifiée de $\mathbf{Q}_p$ de degré h. Dans cette algèbre, on a alors

$$\pi^2 - a_p\pi + p = \pi^2 - a'_p\pi + p = 0,$$

d'où il s'ensuit que $a_p = a'_p$, pour presque tout nombre premier p. Les fonctions L associées à E et à E' sont donc égales (cf. proposition de la page IV.15 de [14]).

4 - Un critère de dépendance algébrique.

Sous les hypothèses équivalentes i) - iii), nous allons mettre en évidence deux réseaux de périodes Λ et Λ', associés respectivement à E et à E', pour lesquels les fonctions de Weierstrass correspondantes $\wp_1$ et $\wp_2$ sont algébriquement dépendantes. Pour cela, D.V. et G.V. Choodnovsky établissent tout d'abord un critère général de dépendance algébrique : le théorème 1.1 de [3], dont nous donnons ici une version simplifiée.

Théorème 5. Soient f et g deux séries formelles à coefficients entiers rationnels, h un isomorphisme analytique local au voisinage de l'origine de $\mathbb{C}$. On suppose que les séries formelles composées $f\circ h$ et $g\circ h$ sont les séries de Taylor en 0 de fonctions méromorphes d'ordre fini (c'est-à-dire : quotient de fonctions entières d'ordre fini). Alors f et g sont algébriquement dépendants sur $\mathbb{Q}$.

Preuve : Il s'agit de mettre en évidence un polynôme P non nul, à coefficients entiers rationnels, tel que $P(f,g)\equiv 0$. En fait, nous allons construire une famille $(P_N)_{N\geq 0}$ de polynômes, telle que P_N convienne pour N suffisamment grand.

Premier pas : construction des P_N.

Pour tout entier $N\geq 10$, il existe un polynôme P_N dans $\mathbb{Z}[X,Y]$, non identiquement nul, de degré $\leq N$ et de hauteur (la hauteur d'un polynôme est le maximum des valeurs absolues de ses coefficients) majorée par $c_1^{N^{3/2}}$, tel que

$$\operatorname{ord}_{t=0} P_N(f(t),g(t)) \geq N^{3/2}.$$

Pour tout polynôme P de $\mathbb{Z}[X,Y]$, de degré $\leq N$, posons

$$R(t) = P(f(t),g(t)).$$

Les équations

$$R(0) = R'(0) = \dots = R^{(k)}(0)/k! = 0$$

avec $k = [N^{3/2}]$, détermine un système de $k+1$ équations linéaires, à coefficients entiers, en les coefficients du polynôme P. Ces derniers sont en nombre

$$(N+1)(N+2)/2 > k+1$$

et le système linéaire ci-dessus admet donc une solution entière non triviale. De plus, les séries formelles f et g sont localement convergentes. On a donc des majorations du type

$$|f^{(n)}(0)/n!| \leq c_2^n , \quad |g^{(n)}(0)/n!| \leq c_2^n ,$$

valable pour tout entier $n \geq 0$. Il est alors facile de vérifier que les coefficients du système linéaire considéré ont une valeur absolue $\leq c_3^k$. Le lemme de Siegel permet de trouver une solution de ce système qui soit au plus du même ordre de grandeur.

<u>Deuxième pas</u> : Si N est suffisamment grand, la série formelle

$$R_N(t) = P_N(f(t), g(t))$$

est identiquement nulle.

Sinon, désignons par n l'ordre en $t=0$ de la série formelle $R_N(t)$. Pour N suffisamment grand, nous allons vérifier l'inégalité

$$|R_N^{(n)}(0)/n!| < 1,$$

qui est manifestement impossible puisque $R_N^{(n)}(0)/n!$ est un entier rationnel, non nul par construction.

Pour cela, faisons le changement de variables

$$t = h(z) = h_1 z + \ldots$$

Par hypothèse, il existe des fonctions entières θ, θ_1, θ_2, d'ordre $\leq \rho$, telles que

$$f \circ h(z) = \theta_1(z)/\theta(z), \quad g \circ h(z) = \theta_2(z)/\theta(z).$$

On peut supposer sans restriction que $\theta(0) \neq 0$. Introduisons alors la fonction entière

$$F_N(z) = \theta(z)^N R_N(h(z)) = \theta(z)^N P_N(f \circ h(z), g \circ h(z)).$$

Puisque R_N a un zéro d'ordre n en $t=0$, on a

$$F_N^{(n)}(0)/n! = \theta(0)^N . h_1^n . R^{(n)}(0)/n! \ .$$

Appliquons maintenant le lemme de Schwarz à la fonction F_N sur un disque de centre 0 et de rayon

$$r = (n/N)^{1/\rho} \geq N^{1/2\rho}$$

(car $n \geq k+1 \geq N^{3/2}$).

On a alors

$$|F_N^{(n)}(0)/n!| \leq |F_N|_r . r^{-n} \leq |F_N|_r . N^{-n/2\rho},$$

où $|F_N|_r$ désigne le maximum de la fonction F_N sur le disque de centre 0 et de rayon r. Puisque les fonctions θ, θ_1, θ_2 sont d'ordre $\leq\rho$, on a la majoration

$$\max(|\theta|_r, |\theta_1|_r, |\theta_2|_r) \leq c_4^{n/N},$$

d'où il s'ensuit que

$$|F_N|_r \leq c_5^{k+n} \leq c_5^{2n}.$$

Nous en déduisons finalement la majoration

$$|R_N^{(n)}(0)/n!| = |\theta(0)|^{-N}. |h_1|^{-n}. |F_N^{(n)}(0)/n!| \leq c_6^n . N^{-n/2\rho}.$$

Cette dernière quantité est <1 si $N > c_6^{2\rho}$.

5 - Preuve du théorème d'isogénie :

Il s'agit de montrer que, si les propriétés équivalentes i) - iii) sont satisfaites, les courbes E et E' sont isogènes sur $\mathbb{Q}$.

Dans la classe d'isomorphisme $\hat{E}=\hat{E}'$, fixons deux lois de groupe formel F et F' telles que F (resp. F') soit algébrique relativement à E (resp. E'). Soit φ un $\mathbb{Z}$-isomorphisme $F \xrightarrow{\sim} F'$ tel que $\varphi(u) = u + \dots$ (on sait qu'il existe un $\mathbb{Z}$-isomorphisme entre F et F'. Si cet isomorphisme est de la forme $\varphi(u) = -u + \dots$, il suffit de le composer avec la symétrie de F ou de F'). Au niveau des exponentielles, on a alors la relation

$$\exp_{F'} = \varphi \circ \exp_F .$$

Appliquons maintenant le critère précédent aux deux séries formelles, à coefficients entiers, x et $\varphi(x)$, avec pour changement de variables $x = \exp_F(z)$. D'après le lemme du paragraphe 2, les séries formelles composées $\exp_F(z)$ et $\exp_{F'}(z)$ se prolongent en des fonctions elliptiques de réseaux Λ et Λ' respectivement. Les hypothèses du critère sont donc satisfaites et il existe une relation algébrique non triviale liant les séries formelles x et $\varphi(x)$. Substituant de nouveau $x=\exp_F(z)$, on obtient une relation polynomiale

$$P(\exp_F(z), \exp_{F'}(z)) \equiv 0,$$

d'où il s'ensuit que les réseaux Λ et Λ' sont commensurables (l'indice de $\Lambda \cap \Lambda'$ dans Λ ou Λ' est << au degré de P). Soit n l'indice de $\Lambda \cap \Lambda'$ dans Λ. La multiplication par $n : \mathbf{C}/\Lambda \longrightarrow \mathbf{C}/\Lambda'$ définit sur les points complexes une isogénie de E dans E'. Comme les invariants des réseaux Λ et Λ' sont rationnels, il est clair que cette isogénie est définie sur $\mathbf{Q}$: si $\wp_1$ et $\wp_2$ désignent les fonctions de Weierstrass associées respectivement aux réseaux Λ et Λ', $\wp_2(nz)$ s'exprime comme fraction rationnelle à coefficients rationnels en $\wp_1(z)$.

6 - <u>Questions d'effectivité</u>.

Le théorème d'isogénie affirme que deux courbes elliptiques définies sur $\mathbf{Q}$, ayant même fonction L, sont isogènes. Rendre effectif ce résultat, c'est déterminer de façon effective, un ensemble fini P de nombres premiers tel que l'égalité des facteurs locaux L_p, p décrivant P, des fonctions L_E et $L_{E'}$, assure l'égalité de ces fonctions. C'est aussi borner de façon effective, le degré d'une isogénie liant alors E et E'. La méthode que nous venons de décrire permet d'obtenir un tel résultat.

Comme précédemment, soit

$$y^2+a_1xy+a_3y= x^3+a_2x^2+a_4x+a_6, \quad a_i\in\mathbf{Z}$$

un modèle de Weierstrass de $E_{\mathbf{Z}}$, et soit Λ le réseau des périodes de la forme différentielle de Néron $dx/(2y+a_1x+a_3)$. Notons

$$H(E)= (\mathrm{vol}(\Lambda)/\pi)^{-1/2},$$

où $\mathrm{vol}(\Lambda)$ désigne le volume euclidien d'une maille du réseau Λ. On notera que $H(E)$ est indépendant du choix d'un modèle de Weierstrass minimal et que $H(E)$ coïncide avec la hauteur (exponentielle) de Faltings, normalisée comme dans [4], lorsque la courbe elliptique $E_{\mathbf{Z}}$ est semi-stable. D.V. et G.V. Choodnovsky énonce alors le théorème suivant.

<u>Théorème 6</u>. Pour tout $\epsilon>0$, il existe une constante C_ϵ telle que, si E et E' désignent des courbes elliptiques, définies sur $\mathbf{Q}$, dont les fonctions L ont les mêmes facteurs locaux pour tout nombre premier

$$p \leq C_\epsilon \max(1,H(E)H(E'))^{2+\epsilon},$$

alors E et E' sont liées par une isogénie de degré inférieur ou égal à

$$C_\epsilon \max(1, H(E)H(E'))^{2+\epsilon}.$$

Nous nous proposons maintenant de comparer ce résultat avec les énoncés antérieurs, dus à J.-P. Serre [13], [5], que voici.

Avec les notations du formulaire de Tate [16], soient

$$\Delta = (c_4^3 - c_6^2)/1728, \quad \Delta' = (c_4^3 - c_6^2)/1728,$$

les discriminants des courbes E et E' respectivement. Notons

$$N_E = \prod_{\ell|\Delta} \ell, \quad N_{E'} = \prod_{\ell|\Delta'} \ell, \quad N = \prod_{\ell|\Delta\Delta'} \ell,$$

(où ℓ désigne toujours un nombre premier).

<u>Théorème 7</u>. Les fonctions L des courbes elliptiques E et E' sont égales si leurs facteurs locaux L_p coïncident pour

i) $p \leq N^c, \quad (p,N)=1,$

où c désigne une constante universelle non explicitée.

ii) $p \ll (\log N)^2, \quad (p,N)=1,$

si l'on admet l'hypothèse de Riemann généralisée.

iii) $p \leq \frac{N}{6} \prod_{\ell|N} (1 + \frac{1}{\ell}),$

si les courbes E et E' sont des courbes de Weil (i.e. : les fonctions L sont des formes modulaires) et si elles sont semi-stables (de telle sorte que N_E et $N_{E'}$ coïncident avec leurs conducteurs respectifs).

Nous allons voir que l'assertion i) de ce théorème est tout-à-fait comparable à la condition du théorème 6, la valeur de la constante c départageant les deux résultats.

Il s'agit donc de comparer N avec H(E)H(E'). Soient (ω_1, ω_2) une base directe de Λ, $\tau = \omega_2/\omega_1$, $(\mathrm{Im}\, \tau > 0)$, et $q = \exp(2\pi i \tau)$. On a alors

$$\Delta = \left(\frac{\omega_1}{2\pi i}\right)^{-12} q \prod (1-q^n)^{24}.$$

Reprenant le calcul de la hauteur de Faltings H(E), effectué par Deligne dans [4], on vérifie que

$$H(E) = |\omega_1^2 \mathrm{Im}\ \tau/\pi|^{-1/2}$$
$$= V(\tau).|\Delta|^{1/12}$$

avec

$$V(\tau) = (4\pi.\mathrm{Im}\ \tau.|q\Pi(1-q^n)^{24}|^{1/6})^{-1/2}.$$

Cette fonction V est invariante sous l'action de $SL_2(\mathbf{Z})$ et peut donc s'exprimer comme fonction de l'invariant modulaire

$$j = c_4^3/\Delta = \frac{1}{q} + 744 + \ldots$$

Il s'ensuit que l'ordre de grandeur de $V(\tau)$ est comparable, à constantes multiplicatives près, à

$$\max(1,|j|)^{1/12}(\log \max(e,|j|))^{-1/2}.$$

On a donc

$$T(1+\log T)^{-1/2} \ll H(E) \ll T,$$

où
$$T = \max(|c_4^3|,|\Delta|)^{1/12}.$$

Remarquons que si la courbe elliptique E est semi-stable, Δ et c_4 sont premiers entre eux et le nombre $\max(|c_4|^3,|\Delta|)$ n'est autre que la hauteur usuelle de l'invariant modulaire j. On a toujours l'inégalité

$$T \geq |\Delta|^{1/12} \geq N_E^{1/12}$$

et donc

$$H(E)H(E') \underset{\epsilon}{\gg} (N_E N_{E'})^{(1-\epsilon)/12} \geq N^{(1-\epsilon)/12}.$$

Il s'ensuit que la condition i) du théorème 7 est plus faible que celle du théorème 6 si la constante c est $<1/6$.

Pour obtenir une comparaison en sens inverse, il nous faut majorer T en fonction de N_E. Il s'agit là d'un problème diophantien non trivial. La théorie des formes linéaires de logarithmes permet d'étudier les facteurs premiers de nombres de la forme $c_4^3 - c_6^2$, mais les résultats obtenus sont trop faibles pour être appliqués utilement. De façon conjecturale, on a cependant l'inégalité suivante :

<u>Conjecture H de</u> [7] : Supposons que c_4^3 et Δ soient premiers entre eux. Alors, pour tout $\epsilon>0$, on a

$$H(j)=\max(|c_4^3|,|\Delta|) \ll_\epsilon N_E^{6+\epsilon}.$$

Cet énoncé est un corollaire simple de la "conjecture abc" de Masser-Oesterlé (cf. [7]). Notons qu'en majorant trivialement N_E par $|\Delta|$, nous retrouvons la conjecture bien connue de M. Hall :

$$|c_4^3-c_6^2| \gg_\epsilon |c_4|^{(1-\epsilon)/2}$$

pour tout couple d'entiers (c_4,c_6) tels que $c_4^3 \neq c_6^2$. Il s'ensuit alors que

$$H(E)H(E')^2 \ll_\epsilon (N_E N_{E'})^{1+\epsilon} \leq N^{2+2\epsilon}$$

Si la constante c est >2, le théorème 6 contient conjecturalement le cas i) du théorème 7.

BIBLIOGRAPHIE

[1] D. Bertrand.- Galois orbits on abelian varieties and zero estimates, L.M.S. Lecture Notes, Vol. 109, 1986.

[2] P. Cartier.- Groupes formels, fonctions automorphes et fonctions zêta des courbes elliptiques, Congrès Int. de Nice, 1970, 291-299.

[3] D.V. et G.V. Choodnovsky.- Padé approximations and diophantine geometry, Proc. Nat. Acad. Sc. USA, vol. 82, 2212-2216.

[4] P. Deligne.- Preuve des conjectures de Tate et Shafarevitch, Sém. Bourbaki, 1983/1984, N° 616.

[5] P. Deligne.- Représentations ℓ-adiques, exposé IX de [15].

[6] G. Faltings.- Eindlichkeitsätze für abelsche Varietäten über Zahlkörpern, Invent. Math. 73, 1983, 349-366.

[7] G. Frey.- Links between elliptic curves and solutions of A-B=C, à paraître.

[8] W. Hill.- Formal groups and zeta-functions of elliptic curves, Invent. Math. 12, 1971, 321-336.

[9] T. Honda.- On the theory of commutative formal groups, J. Math. Soc. Japan, vol. 22, 1970, 213-246.

[10] S. Lang.- Introduction to transcendental numbers, Addison-Wesley, 1966.

[11] S. Lang.- Division points of elliptic curves and abelian functions over number fields, Am. J. of Math. vol. 97, 1975, 124-132.

[12] D. Masser.- Division fields of elliptic functions, Bull. London Math. Soc. 9, 1977, 49-53.

[13] J.-P. Serre.- Quelques applications du théorème de densité de Cebotarev. Pub. Math. IHES, 54, 1981, 123-202.

[14] J.-P. Serre.- Abelian ℓ-adic representations and elliptic curves, Benjamin, New-York, 1968.

[15] L. Spiro.- Séminaire sur les pinceaux arithmétiques : la conjecture de Mordell, Astérisque, vol. 127, 1985.

[16] J. Tate.- The arithmetic of elliptic curves, Invent. Math. 23, 1974, 179-206.

Michel LAURENT
I.H.P.
11, rue P. et M. Curie
75231 PARIS CEDEX 05

On Quadratic Gauss Sums over Local Fields

Wen-Ch'ing Winnie Li*

1. Introduction.

The classical Gauss sum attached to the Legendre symbol $(\frac{\cdot}{p})$ with an odd prime p is

$$g_p = \frac{1}{\sqrt{p}} \sum_{\substack{x \bmod p \\ (x,p)=1}} (\frac{x}{p}) e^{2\pi i x/p} = \frac{1}{\sqrt{p}} \sum_{x \bmod p} e^{2\pi i x^2/p},$$

which is equal to 1 for $p \equiv 1 \pmod 4$ and i for $p \equiv 3 \pmod 4$. At any rate, it is a fourth root of 1. Denote by ψ the additive character on $\mathbb{Z}/p\mathbb{Z}$ given by $\psi(y) = e^{2\pi i y/p}$, then we can rewrite the above sum as

$$g_p = \frac{1}{\sqrt{p}} \sum_{\substack{x \bmod p \\ (x,p)=1}} (\frac{x}{p}) \psi(x) = \frac{1}{\sqrt{p}} \sum_{x \bmod p} \psi(x^2).$$

Here the first expression can be regarded as the Fourier transform of the multiplicative character $\chi(x) = (\frac{x}{p})$ of $(\mathbb{Z}/p\mathbb{Z})^\times$ with respect to the additive character ψ, while the second is the sum over all elements of $\mathbb{Z}/p\mathbb{Z}$ of ψ composed with the quadratic form $Q(x)=x^2$. In this article the base field F, in lieu of the finite field $\mathbb{Z}/p\mathbb{Z}$, will be a nonarchimedean local field with q elements in its residue field. Let ψ be a nontrivial additive character of F and χ a multiplicative character of $F^\times$. The role of Gauss sum over F, in analogy with g_p, is played by

$$\gamma^F(\chi,\psi) = \int_{F^\times} \chi(x)\psi(x)\, d_\psi^* x,$$

where $d_\psi^* x = |x|^{-1/2} d_\psi x$, $|\ |$ is the usual absolute value on F, and $d_\psi x$ is the Haar measure on F self-dual with respect to ψ. The integral is taken in principal value, which converges if χ is ramified or if $\chi = Z^{\mathrm{ord}}$ is unramified for some complex number Z with absolute value less than $q^{1/2}$. As the group of characters of $F^\times$, $A(F^\times)$, has an analytic structure with each connected component consisting of χZ^{ord}, $Z \in \mathbb{C}^\times$, and the integral converges to a rational function on each component of $A(F^\times)$, we extend $\gamma^F(\chi,\psi)$ to all characters χ by analytic continuation. We are interested in two kinds of "quadratic" Gauss sums stemmed from the above two interpretations of g_p.

The first kind has the form $\int_V \psi \circ Q(x) d_{\psi\circ T} x$ philosophically, where Q is a nondegenerate quadratic form on a finite-dimensional vector space V over F, T is the bilinear form induced from Q :

$$T(x,y) = Q(x+y) - Q(x) - Q(y),$$

and the Haar measure on V is chosen to be self-dual with respect to the pairing $(x,y) \longmapsto \psi\circ T(x,y)$. Such an integral, if converges, has values equal to an 8th root of 1, as studied by Weil in [W]. See also Gérardin [G]. We will be concerned with the special case where Q is the reduced norm N_A over F of a rank two separable semi-simple F-algegra A. In this case the integral, noted by $\lambda_A(\psi)$, is a 4th root of 1. More details will be reviewed in section 2.

The second kind is the integral of a smooth irreducible representation π of the unit group $A^\times$ of a rank two separable semi-simple F-algebra A against ψ composed with the reduced trace T_A of A over F. It is in fact a scalar operator (sometimes to be interpreted distributionally) which we shall identify with the scalar. Since the conjugacy classes of $A^\times$ are parametrized by the map (N_A, T_A) to $F^\times \times F$, in order to see the integral as a Fourier transform of π, we twist π by a character χ of $F^\times$ composed with N_A and multiply the resulting integral by $\lambda_A(\psi)$; in other words, we view

$$\gamma_\pi(\chi,\psi) = \lambda_A(\psi) \int_{A^\times} \pi(x)\chi(N_A x)\psi(T_A x)\, d^*_{\psi\circ T_A} x$$

as a quadratic Gauss sum. Here the integral is taken in principal value and extended to all χ by analytic continuation, $d^*_{\psi\circ T_A} x = |N_A x|^{-r/2} d_{\psi\circ T_A} x$, $d_{\psi\circ T_A} x$ is the Haar measure on A self-dual with respect to $\psi\circ T_A$, and $r=1$ if A is commutative, and $r=2$ otherwise. Note that when π is finite-dimensional, one may replace π by its reduced trace $c_\pi = \mathrm{tr}\, \pi/\deg \pi$ so that the integral definig γ_π is actually a scalar. As general as these sums γ_π may appear to be, we will see later that the "majority" of them are of the form

$\gamma^F(\mu\chi,\psi)\gamma^F(\nu\chi,\psi)$ arising from a character $\pi = (\mu,\nu)$ of $(F\times F)^\times$ or $\gamma^F(\eta_K,\psi)\gamma^K(\theta\bullet\chi\circ N_K,\psi\circ T_K)$ arising from a multiplicative character $\pi=\theta$ of a separable quadratic extension K of F. Here η_K is the quadratic character of $F^\times$ attached to K. Our objective is to characterize the quadratic Gauss sums γ_π, regarded as functions in χ and ψ, in terms of certain identities they satisfy. This was begun with my paper [L 1] and continued in a series of joint work with P. Gérardin [GL 1-6]. Some results were reported in our article [L 2]; the criteria there were the simplest possible except for the case where A is a separable quadratic extension of F, which is the subject of this article. For the sake of completeness, we summarize below our criteria for all A.

Recall that a rank two separable semi-simple F-algebra A is isomorphic to (a) F×F, or (b) a separable quadratic extension K of F, or (c) the algebra $M_2(F)$ of 2x2 matrices over F, or (d) a quaternion algebra H over F. For any A as above and any smooth irreducible representation π of $A^\times$, the function $\gamma(\chi,\psi)=\gamma_\pi(\chi,\psi)$ is a rational function in χ (with ψ fixed) and satisfies the following two identities (cf. Theorem 3.3 of [L2]) :

(AC) $$\gamma(\chi,\psi^t)= (\omega^{-1}\chi^{-2})(t)\gamma(\chi,\psi) \quad \text{for all } t \text{ in } F^\times,$$

where ψ^t is the additive character of F defined by $\psi^t(u)= \psi(tu)$, $u\in F$, and ω is the restriction of π to $F^\times$ (embedded in $A^\times$) except when A is of type (b) in which case the restriction of π to $F^\times$ equals $\omega\eta_K$;

(MF) $$\int_C \gamma(\alpha\chi^{-1},\psi)\Gamma(\beta\chi^{-1},\psi)\gamma(\chi,\psi)d\chi= \gamma(\alpha,\psi)\gamma(\beta,\psi)\Gamma(\alpha^{-1}\beta^{-1}\omega^{-1},\psi^{-1}),$$

where

$$\Gamma(\chi,\psi)= \int_{F^\times} \chi(x)\psi(x)d^\times x$$

with the Haar measure $d^\times x$ on $F^\times$ normalized so that the group of units $\mathcal{O}^\times$ of $F^\times$ has volume 1, $d\chi$ on $A(F^\times)$ is equal to the counting measure on the dual of $\mathcal{O}^\times$ times $\frac{1}{2\pi i}\frac{dZ}{Z}$ on each connected component. C is the union of simple positively oriented contours on each connected component of $A(F^\times)$ enclosing the origin, but not the poles of γ, the integral is in principal value, and the identity is valid for α, β enclosed by C so that $\alpha\beta\omega$ has absolute value >1 and then extended by analytic continuation.

It turns out that these two are the main identities in our criteria.

Theorem 1.1. let $\gamma(\chi,\psi)$ be a function rational in χ.

(a) $\gamma=\gamma_\pi$ for a character π of $F^\times \times F^\times$ if and only if γ satisfies (AC), (MF) and γ has at least one pole. Moreover, the poles of γ determine the pair $\{\mu,\nu\}$ of characters of $F^\times$ such that $\pi=(\mu,\nu)$; then $\gamma(\chi,\psi)=\gamma^F(\mu\chi,\psi)\gamma^F(\nu\chi,\psi)$.

(b) $\gamma=\gamma_\theta$ for a character θ of $K^\times$ if and only if γ satisfies (AC), (MF) and

(K-type) $\quad \gamma(\chi\eta_K,\psi)=\gamma(\chi,\psi)$ for all characters χ.

Further, γ determines θ up to conjugation by Gal(K/F).

(c) $\gamma=\gamma_\pi$ for an infinite-dimensional admissible irreducible representation π of $GL_2(F)$ if and only if γ satisfies (AC) and (MF). Moreover, γ determines the class of π.

(d) $\gamma=\gamma_\rho$ for an irreducible representation ρ of $H^\times$ if and only if γ satisfies (AC), (MF) and γ has at most one pole. Furthermore, γ determines the reduced character and hence the class of ρ.

This theorem yields immediately the correspondences between classes of representations of different unit groups $A^\times$ so that the corresponding classes have the same associated γ-factor, as discussed in [L2].

To each degree two semi-simple representation σ of the Weil group W_F over F we define γ_σ using the associated L- and ϵ-factors :

$$\gamma_\sigma(\chi,\psi)=\frac{\epsilon(1/2,\check{\sigma}\otimes\chi^{-1},\psi)L(1/2,\sigma\otimes\chi)}{L(1/2,\check{\sigma}\otimes\chi^{-1})},$$

where $\check{\sigma}$ is the contragredient of σ. One refers to [D] for the definition of L- and ϵ-factors. When σ is induced from a multiplicative character θ of a separable quadratic extension K of F, the induction properties of the L- and ϵ-factors attached to σ imply that $\gamma_\sigma=\gamma_\theta$. Thus the statement (b) in the theorem above has the following alternative assertion :

Theorem 1.2. A function $\gamma(\chi,\psi)$ rational in χ is equal to γ_σ for some degree 2 induced representation σ of W_F if and only if γ satisfies (AC), (MF) and (K-type) for some separable quadratic extension K of F.

Combined with Theorem 1.1, (c), this yields immediately

<u>Corollary 1.3</u> (Langlands). An infinite-dimensional admissible irreducible representation π of $GL_2(F)$ corresponds to an induced degree two representation of W_F such that the corresponding representation have the same attached γ-factor if and only if π is equivalent to $\pi \otimes \eta_K$ for some separable quadratic extension K of F.

This result, according to Langlands, played a critical role in his theory of base change for GL(2); he was not satisfied with his proof given in [LA], Lemma 7.17, where he computed the character of a virtual representation arising from restricting π to a subgroup of $GL_2(F)$ of index 2. We hope our proof is more satisfactory.

When the residual characteristic of F is odd, one can show that any supercuspidal representation π of $GL_2(F)$ is equivalent to $\pi \otimes \eta_K$ for some K (cf. GL[6]), and in this case all irreducible degree two semi-simple representations of W_F are induced, thus one has

<u>Corollary 1.4</u>. Suppose the residual characteristic of F is odd. Then there is a bijection between equivalence classes of supercuspidal representations of $GL_2(F)$ and equivalence classes of degree two semi-simple irreducible representations of W_F such that the corresponding classes have the same attached γ-factor.

Since the γ-factor attached to an irreducible representation ρ of $H^\times$ has a pole if and only if ρ is one-dimensional, in view of Theorem 1.1, (c), (d), we may replace the classes of supercuspidal representations of $GL_2(F)$ in Corollary 1.4 by the classes of irreducible representations of $H^\times$ of degree >1.

The last two sections concern the proof of Theorem 1.1, (b). In section 3 we list properties satisfied by γ_θ, including (AC), (MF) and (K-type), and thereby establish the necessity. The proof of sufficiency is sketched in section 4, details will appear in [GL5-6]. Given γ as described, we shall perform Fourier inversion on $F^\times \times F$ to define $c_\theta (=\frac{1}{2}(\theta+\bar{\theta}))$. The main difficulty lies in showing that the inverted function on $F^\times \times F$ vanishes outside the image of $K^\times$ under the map (N_K, T_K). This is partially equivalent to the statement that for any separable quadratic extension $K' \neq K$, the inner product of γ and $\gamma_{\theta'}$ is independent of the choice of the character θ' of $K'^\times$ with prescribed restriction to $F^\times$. The value of this inner product is given in Theorem 4.4. One may also regard the identity there as an explicit formula for the γ-factor attached to the representation $\pi\times\pi'$ of $GL_2(F)\times GL_2(F)$, where $\gamma=\gamma_\pi$ and $\gamma_{\theta'}=\gamma_{\pi'}$.

2. <u>The Gauss sum</u> $\lambda_A(\psi)$.

Let A be a rank two separable semi-simple F-algebra with reduced norm N_A and reduced trace T_A over F. Let ψ be a non trivial additive character of F. Write $\underline{g}$ for $\psi \circ N_A$. The Gauss sum $\lambda_A(\psi)$ occurs in the functional equation :

$$\int_A \hat{f}(x)\underline{g}(x)d_{\psi\circ T_A}x = \lambda_A(\psi) \int_A f(x)\underline{g}(x)^{-1}d_{\psi\circ T_A}x ,$$

where f is any Schwartz function on A and $\hat{f}$ is the Fourier transform of f with respect to $\psi \circ T_A$. Thus its value may be computed by choosing an appropriate f and calculating the quotient. The result is given in [W] (see also [G], [GL 1] and [GL 5]) :

$$\begin{aligned}\lambda_A(\psi) &= 1 && \text{if } A \cong F \times F \text{ or } M_2(F),\\ &= -1 && \text{if } A \cong H,\\ &= \gamma^F(\eta_K,\psi) && \text{if } A \cong K, \text{ a separable quadratic extension of } F.\end{aligned}$$

In the last case, it follows from the functional equation

$$\lambda_K(\psi)\lambda_K(\psi^{-1}) = \lambda_K(\psi)^2\eta_K(-1)=1$$

that $\lambda_K(\psi)^2 = \eta_K(-1) = \pm 1$. Therefore $\lambda_A(\psi)$ is a fourth root of 1 in all cases. Note that if we take f to be the characteristic function supported on a neighborhood of 0 in A divided by the volume of this neighborhood and shrink this neighborhood to the point 0, then the Fourier transform of f tends to the characteristic function of A; this process yields the formal expression given in the Introduction :

$$\lambda_A(\psi) = \int_A \psi(N_A x)\ d_{\psi\circ T_A}\ x.$$

When A is non-split, that is, A is isomorphic to H or K, the integral on the right hand side converges in principal value to $\lambda_A(\psi)$, as computed in [W], [GL 1] and [GL 5]. When A splits, although the integral is no longer convergent in principal value, it still serves as a good philosophical ground. For instance, in case $A = F\times F$, this integral formally equals

$$\int_{F\times F} \psi(xy)d_\psi\ x\ d_\psi\ y = \int_F \hat{1}(y)d_\psi y = \hat{1}(0)=1.$$

3. Properties satisfied by γ_θ.

In what follows we fix a separable quadratic extension K of F. We study the properties satisfied by the function γ_θ attached to a character θ of $K^\times$. Recall that, for a character χ of $F^\times$ and a non-trivial character ψ of F,

$$\gamma_\theta(\chi,\psi)= \lambda_K(\psi)\, \gamma^K(\theta.\chi\circ N_K,\ \psi\circ T_K)$$

$$= \lambda_K(\psi) \int_{K^\times} \theta(x)\chi(N_K x)\psi(T_K x)d^*_{\psi\circ T_K} x$$

$$= \int_{F^\times} \eta_K(u)\psi(u)d^*_\psi u \int_{K^\times} c_\theta(x)\ \chi(N_K x)\psi(T_K x)d^*_{\psi\circ T_K} x,$$

where $c_\theta = \frac{1}{2}(\theta+\overline{\theta})$ and $\overline{\theta}$ is the image of θ under the Galois conjugation given by $\overline{\theta}(x)= \theta(\overline{x})$ with $x \mapsto \overline{x}$ being the generator of Gal(K/F). This integral expression of γ_θ yields immediately the following

Proposition 3.1. Write the restriction of θ to $F^\times$ as $\omega\eta_K$. Then $\gamma=\gamma_\theta$ satisfies the conditions (AC) and (K-type) (as in Introduction).

Proposition 3.2 (Davenport-Hasse identity). If $\theta= \mu\circ N_K$ for some character μ of $F^\times$, then $\gamma_\theta(\chi,\psi)= \gamma^F(\mu\chi,\psi)\gamma^F(\mu\eta_K\chi,\psi)$.

This can be seen either from representations of $GL_2(F)$ as discussed in [L 1] or by a direct computation as in [GL 5].

Proposition 3.3. Let c be a locally constant function on $K^\times$ with $c(1)=1$, invariant under Gal(K/F), and of type $\omega\eta_K$, that is,

$$c(ux)= (\omega\eta_K)(u)c(x) \quad \text{for } u\in F^\times \text{ and } x\in K^\times.$$

Define the function $\gamma(\chi,\psi)$ in characters χ of $F^\times$ and non trivial additive characters ψ of F by the following integral and its analytic continuation :

$$\gamma(\chi,\psi)= \lambda_K(\psi)\int_{K^\times} c(x)\chi(N_K x)\psi(T_K x)d^*_{\psi\circ T_K} x.$$

(It is a rational function in χ). Then c satisfies

(3.4) $\qquad \frac{1}{2}(c(xy)+c(x\overline{y}))= c(x)c(y)$ for x,y in $K^\times$

if and only if γ satisfies (MF) with the same ω.

Since c is a function on $K^{\times}$ modulo the action of Gal(K/F), which can be identified with the image of $K^{\times}$ under the map (N_K, T_K), we view both sides of (3.4) as functions on the product of the image $(N_K, T_K)(K^{\times})$ with itself. Then (MF) is nothing but the Fourier transform of the identity (3.4). As the function $c=c_{\theta}$ satisfies the assumptions in Proposition 3.3 and also (3.4), consequently, we have

<u>Corollary 3.5</u>. γ_{θ} satisfies (MF).

In particular, we have established the necessity of Theorem 1.1,(b).

Let γ and γ' be two functions in χ and ψ satisfying (AC) and (MF) with characters ω and ω', respectively. Given a non trivial additive character ψ of F, we define the "inner product" of γ and γ' to be the finite part of the integral below :

$$\langle\gamma,\gamma'\rangle_{\psi}=\int_C \gamma(\chi,\psi)\gamma'(\chi^{-1},\psi^{-1})d\chi,$$

where C is the union of contours, one on each connected component of $A(F^{\times})$, enclosing 0 and the poles of $\gamma'(\chi^{-1},\psi^{-1})$ but not the poles of $\gamma(\chi,\psi)$. In [GL 5] the following identity is proved.

<u>Theorem 3.6</u>. Let K, K' be two distinct separable quadratic extensions in an algebraic closure of F. Write B for the biquadratic extension KK'. Let θ (resp. θ') be a character of $K^{\times}$ (resp. $K'^{\times}$) with restriction to $F^{\times}$ equal to $\omega\eta_K$ (resp. $\omega'\eta_{K'}$). Then

$$\langle\gamma_{\theta},\gamma_{\theta'}\rangle_{\psi}=\frac{\lambda_B}{\Gamma(\omega\omega' q^{-2\mathrm{ord}},\psi)}\,\gamma^{B}(\xi\cdot q^{-1/2\mathrm{ord}_B},\psi\circ T_B),$$

where $\xi=(\theta\circ N_{B/K})(\theta'\circ N_{B/K'})$ and $\lambda_B=\prod_{\substack{B\supset E\supset F\\ [E:F]=2}}\lambda_E(\psi)$ is independent of ψ.

<u>Corollary 3.7</u>. If in Theorem 3.6 we have $\omega\eta_K\omega'\eta_{K'}=q^{\mathrm{ord}}$, then

$$\langle\gamma_{\theta},\gamma_{\theta'}\rangle_{\psi}=\lambda_{B/K}/\Gamma(\eta_K\eta_{K'}q^{-\mathrm{ord}},\psi^{-1}).$$

Note that the value of $\langle\gamma_{\theta},\gamma_{\theta'}\rangle_{\psi}$ above depends only on the fields K, K' and the product of the restrictions of θ, θ' to $F^{\times}$. It is this property which will be used in next section to obtain our criterion. Here the value $\lambda_{B/K}$ is independent of ψ; in fact, it is equal to 1 if the characteristic of F is 2, otherwise it is the Hilbert symbol attached to the two quadratic extensions K', K" of F in B other than K. More precisely, if $K' = F(\sqrt{a'})$ and $K'' = F(\sqrt{a''})$ with a', $a'' \in F^\times$, then $\lambda_{B/K} = (a', a'')$. It can also be interpreted as the cup product of $\eta_{K'}$ and $\eta_{K''}$. See Serre [S], Chap. 14, for more detail.

4. Characterizing γ_θ.

We regard $\gamma_\theta(\chi,\psi)$ as a function in χ and ψ and seek to characterize it in terms of certain identities it satisfies. The answer is stated in Theorem 1.1,(b), which be recall :

Theorem 4.1. Let K be a separable quadratic extension of F. Let $\gamma(\chi,\psi)$ be a function in χ, ψ and rational in χ. Then $\gamma=\gamma_\theta$ for some character θ of $K^\times$ if and only if γ satisfies (AC),(MF) and (K-type).

The necessity was proved in the previous section. Now we prove sufficiency. Let $\gamma(\chi,\psi)$ be a function rational in χ and satisfying (AC), (MF) and (K-type). It is shown in [GL 2] that, as a consequence of (AC) and (MF), γ has at most two poles, and if it has at least one pole, then it is equal to $\gamma^F(\mu\chi,\psi)\gamma^F(\nu\chi,\psi)$ for two characters μ,ν of $F^\times$. Thus if our γ has at least one pole, then the (K-type) condition implies

$$\gamma(\chi,\psi) = \gamma^F(\mu\chi,\psi)\gamma^F(\mu\eta_K\chi,\psi),$$

which is equal to $\gamma_{\mu\circ N_K}(\chi,\psi)$ by the Davenport-Hasse identity (Proposition 3.2). Hence the theorem is proved for this case, and we may assume from now on that γ has no pole on $A(F^\times)$. This assumption is not necessary in the argument below, but it will simplify our analysis, in particular, on convergence of integrals.

Recall that for a character θ of $K^\times$, $c_\theta = \frac{1}{2}(\theta+\bar\theta)$ is a function on the Gal(K/F) orbits on $K^\times$, which are parametrized by the map $(N_K, T_K) : K^\times \longrightarrow F^\times \times F$. Regard c_θ as a function on $F^\times\times F$, zero outside the image of $K^\times$ under (N_K, T_K). When we pass the integral for γ_θ from $K^\times$ to $F^\times\times F$, the measure $d^*_{\psi\circ T_K}x$ becomes

$J(x)|\varphi|d^{\times}\nu d\tau$, where the jacobian $J(x)$, also written as $J(N_K x, T_K x)$, equals $|N_K(1-\frac{x}{\bar{x}})|^{1/2}$ on $K-F$, and $|\varphi|=q^{-\mathrm{ord}\,\varphi}$; thus γ_θ can be expressed as

$$\gamma_\theta(\chi,\varphi)=\lambda_K(\varphi)\int_{F^\times\times F} c_\theta(\nu,\tau)\chi(\nu)\varphi(\tau)|\varphi|J(\nu,\tau)d^{\times}\nu\, d\tau. \tag{4.2}$$

Starting with the given γ, we shall first construct a function c on $K^\times/\mathrm{Gal}(K/F)$, then show that $c=c_\theta$ for some character θ of $K^\times$. In view of (4.2), c may be defined by taking the Fourier inversion of (4.2) with γ replacing γ_θ. This is what we shall do. To facilitate our proof, we introduce inversion in a more general context.

Let E be a commutative rank two separable F-algebra. If E splits it is isomorphic to $F\times F$; if it is nonsplit, it is isomorphic to a separable quadratic extension K' of F. For $x\in E^\times$ define

$$S(x)=\int_{\hat{F}}\int_C \gamma(\chi,\varphi)\lambda_K(\varphi)^{-1}\chi^{-1}(N_E x)\varphi(-T_E x)|\varphi|^{-1}d\chi d\varphi,$$

where $\hat{F}$ is the group of all additive characters of F, and C is the union of contours, one on each connected component of $A(F^\times)$ enclosing 0 positively. For $x\in K^\times$, define

$$c(x)=\frac{S(x)}{J(x)}.$$

Rigorously speaking, $c(x)$ is defined by the formula above for x not in $F^\times$; for $x\in F^\times$, we compute the jacobian in a small neighborhood of x and replace $S(x)$ by a suitably truncated integral, as we shrink the neighborhood to the point x, the ratio of modified S by J tends to $c(x)$.

<u>Theorem 4.3</u>. Let c be the function on $K^\times$ defined above. Then

(1) c is a locally constant function with $c(1)=1$, invariant under $\mathrm{Gal}(K/F)$ and of type $\omega\eta_K$.

(2) $$\gamma(\chi,\varphi)=\lambda_K(\varphi)\int_{K^\times} c(x)\chi(N_K x)\varphi(T_K x)d^*_{\varphi\circ T_K}x.$$

<u>Proof</u>. The first assertion basically follows from the definition of c and the condition (AC). The second assertion is where the main difficulty lies because c is defined by taking Fourier inversion on $F^\times\times F$, not on $K^\times$. Note that $F^\times\times F$ is the union of $(N_E,T_E)(E^\times)$ over all commutative, rank two separable F-algebras E when the characteristic of F is not equal to 2; in case the characteristic

of F is equal to 2, the complement of this union in $F^{\times}\times F$ comes from the image of inseparable quadratic extensions of F under the norm and the trace map over F, and hence is contained in $F^{\times}\times\{0\}$, which has measure zero in $F^{\times}\times F$. Therefore (2) is equivalent to

$$(2)' \qquad S(x)=0 \quad \text{for} \quad x\in E-F \quad \text{and} \quad E \neq K.$$

To prove this, we express $S(x)$ as an integral involving the inner product of γ with $\gamma_{\theta'}$ as defined in section 3. More precisely, one can show, for $x\in E-F$, that

$$\int_{C(E)}\langle\gamma,\gamma_{\theta'}\rangle_{\psi}\ \theta'(x)^{-1}d\theta' = \lambda_E(\psi^{-1})\lambda_K(\psi)\,|N_E x|^{1/2}\ \delta_E\ S(x),$$

where

$$\delta_E = q^{-d_E/2}(1-q_E^{-1})(1-q^{-1})^{-1} \quad \text{if } E \text{ is nonsplit},$$

$$= 1-q^{-1} \qquad \text{if } E \text{ splits},$$

d_E is the (exponent of the) conductor of η_E, q_E is the cardinality of the residue field of E and $C(E)$ is the set of characters (resp. the union of contours on components of characters) of $E^{\times}$ whose restriction to $F^{\times}$ is equal to $\omega^{-1}\eta_K\, q^{\mathrm{ord}}$ if E is nonsplit (resp. if E splits). Thus (2)' is equivalent to

$$(2)'' \qquad \int_{C(E)}\langle\gamma,\gamma_{\theta'}\rangle_{\psi}\ \theta'(x)^{-1}d\theta'=0 \quad \text{for} \quad x\in E-F \quad \text{and} \quad E \neq K,$$

which, in turn, is equivalent to

$$(2)''' \qquad \langle\gamma,\gamma_{\theta'}\rangle_{\psi} \ \text{ is independent of } \theta' \text{ in } C(E) \text{ for } E \neq K.$$

When E splits, (2)"' follows from (MF); when E is nonsplit, (2)"' is a consequence of the following theorem. This will conclude the proof of Theorem 4.3.

<u>Theorem 4.4</u>. Let K' be a separable quadratic extension of F different from K. Denote by B the biquadratic extension KK'. For any character θ' of K' whose restriction to $F^{\times}$ equals $\omega^{-1}\eta_K\, q^{\mathrm{ord}}$, we have

$$\langle\gamma,\gamma_{\theta'}\rangle_{\psi}= \lambda_{B/K}/\Gamma(\eta_K\eta_{K'}q^{-\mathrm{ord}},\psi^{-1}).$$

We sketch a proof of this theorem. Applying (MF) to γ and $\gamma_{\theta'}$ we first show that

$$\langle\gamma,\gamma_{\theta'}\rangle_{\psi}=\epsilon(\gamma,\gamma_{\theta'})/\Gamma(\eta_K\eta_{K'}q^{-\mathrm{ord}},\psi^{-1}),$$

where $\epsilon(\gamma,\gamma_{\theta'})$ is a sign. To determine this sign consider the integral of $\langle\gamma,\gamma_{\theta'}\rangle_{\psi}$ over $C_n(K')$, the set of characters in $C(K')$ with conductor $\leq n$. It is equal to

$$\sum_{\theta'\in C_n(K')}\epsilon(\gamma,\gamma_{\theta'})/\Gamma(\eta_K\eta_{K'}q^{-\mathrm{ord}},\psi^{-1})$$

on one hand. On the other hand, one can show that for n large, the integral depends only on K, K' and is independent of γ, $\gamma_{\theta'}$. Thus it is also equal to the integral of $\langle\gamma_{\theta},\gamma_{\theta'}\rangle_{\psi}$ over $C_n(K')$ for some character θ of $K^{\times}$ with conductor $\leq n/2$ and whose restriction to $F^{\times}$ is equal to $\omega\eta_K$. Therefore we have

$$\sum_{\theta'\in C_n(K')}\epsilon(\gamma,\gamma_{\theta'})=\sum_{\theta'\in C_n(K')}\epsilon(\gamma_{\theta},\gamma_{\theta'})=\lambda_{B/K}\cdot\text{the cardinality of } C_n(K')$$

by Corollary 3.7. As $\epsilon(\gamma,\gamma_{\theta'})=\pm1$, this implies

$$\epsilon(\gamma,\gamma_{\theta'})=\lambda_{B/K},$$

as desired.

Now back to the proof of Theorem 4.1. We see from Theorem 4.3 that the function c constructed from γ satisfies the assumptions of Proposition 3.3, thus the (MF) condition implies that c satisfies (3.4), which in turn implies $c=c_{\theta}$ for some character θ of $K^{\times}$. This proves $\gamma=\gamma_{\theta}$ by Theorem 4.3,(2). The proof of Theorem 4.1 is completed.

*p. 133 : Research supported in part by N.S.F grant DMS-8404083 and in part by C.N.R.S.

BIBLIOGRAPHY

[D] P. Deligne.- Les constantes des équations fonctionnelles des fonctions L, in Modular Functions in One Variable II, Lecture Notes in Mathematics 349, Springer-Verlag, Berlin-Heidelberg-New York (1973), 501-597.

[G] P. Gérardin.- Groupes quadratiques et applications arithmétiques, Séminaire de Théorie des Nombres, Paris 1984-85 (Goldstein, ed.), Progress in Math. 63, Birkhäuser Boston (1986).

[GL 1] P. Gérardin and W.-C.W. Li.- Fourier transforms of representations of quaternions, J. reine angew. Math. 359 (1985), 121-173.

[GL 2] P. Gérardin and W.-C.W. Li.- A functional equation for degree two local factors, Canadian Math. Soc. Bulletin 28 (3) (1985), 355-371.

[GL 3] P. Gérardin and W.-C.W. Li.- Establishing correspondences without trace formula, AMS Contemporary Math. 53 (1986), 185-200.

[GL 4] P. Gérardin and W.-C.W. Li.- Identities on degree two local factors. Proceedings of the conference on "Representation Theory and Number Theory in connection with Local Langlands Conjecture", held at Irsee, Germany, Dec. 15-20, 1985 (to appear).

[GL 5] P. Gérardin and W.-C.W. Li.- Identities on quadratic Gauss sums (submitted).

[GL 6] P. Gérardin and W.-C.W. Li.- Degree two monomial representations of local Weil groups (submitted).

[La] R.P. Langlands.- Base Change for GL(2), Ann. Math. Studies 96, Princeton Univ. Press (1980).

[L 1] W.-C.W. Li.- Barnes' identities and representations of GL_2, Part II : Nonarchimedean local field case, J. reine angew. Math. 345 (1983), 69-92.

[L 2] W.-C.W. Li.- Various aspects of Barnes' identity, Séminaire de Théorie des Nombres, Paris 1984-85 (Goldstein ed.), Progress in Math. 63, Birkhäuser Boston (1986), 187-203.

[S] J.-P. Serre.- Corps locaux, Hermann, Paris (1968).

[W] A. Weil.- Sur certains groupes d'opérateurs unitaires, Acta Math. III (1964), 143-211.

Wen-Ch'ing Winnie Li
Department of Mathematics
Pennsylvania State University
University Park, PA 16802
U.S.A.

Points entiers des variétés arithmétiques

Laurent MORET-BAILLY*

1 - Introduction et notations.

1.1. En 1934, Skolem [5] a étudié la question suivante : si $Q\in R[X_1,\ldots,X_m]$ est un polynôme à coefficients dans un anneau R d'entiers algébriques, peut-on trouver des entiers algébriques $x_1,\ldots,x_m$ tels que $Q(x_1,\ldots,x_m)$ soit un entier algébrique <u>inversible</u> ?

Si une solution existe, alors Q est nécessairement <u>primitif</u> (l'idéal de R engendré par ses coefficients est R). Le théorème prouvé par Skolem est que cette condition est aussi suffisante.

1.2. C'est là un cas très particulier du problème suivant : on considère un anneau de Dedekind R de corps des fractions K; on fixe une clôture algébrique K^{alg} de K, et on désigne par R^{int} la fermeture intégrale de R dans K^{alg}. Etant donné un R-schéma séparé de type fini

$$f : X \longrightarrow B = \mathrm{Spec}\, R,$$

a-t-on $X(R^{int}) \neq \emptyset$?

Dans le cas du problème de Skolem, X est l'ouvert de l'espace affine $\mathbb{A}^m_R = \mathrm{Spec}\, R[X_1,\ldots,X_m]$, complémentaire du fermé d'équation $Q=0$; autrement dit, $X = \mathrm{Spec}\,(R[X_1,\ldots,X_m,Y]/(YQ-1))$. La condition "$Q$ primitif" signifie que pour tout idéal premier $P \in \mathrm{Spec}\, R$, on a $Q\not\equiv 0 \pmod P$; en d'autres termes, $f : X \longrightarrow B$ est surjectif.

1.3. On se propose dans cet exposé d'énoncer, et de démontrer succinctement, un théorème très général, dû à R. Rumely, d'existence de solutions pour des problèmes de ce type (théorème 1). A vrai dire, comme on va le voir, l'adjectif "général" porte sur les conditions imposées au morphisme f, l'anneau de base R étant par contre soumis à de sérieuses restrictions, heureusement satisfaites par l'anneau des entiers d'un corps de nombres, ou l'anneau d'une courbe affine sur un corps fini.

2 - Points entiers.

2.1. Nous gardons les notations de 1.2. Si $x : \mathrm{Spec}\,(R^{int}) \longrightarrow X$ est un point de $X(R^{int})$, son image Y dans X est un fermé irréductible de X, surjectif et fini (i.e. propre à fibres finies) sur B. Inversement, un tel Y est de la forme $\mathrm{Spec}\,(R')$, où R' est une R-algèbre intègre contenant R et entière sur R, donc une sous-algèbre de R^{int}, d'où un point de $X(R^{int})$, d'image Y. Nous poserons donc :

<u>Définition 1</u>. <u>Un point entier</u> de X est un fermé irréductible de X, fini et surjectif sur B.

2.2. Si X admet un point entier Y, celui-ci est contenu dans l'une des composantes irréductibles de X; par suite, X doit vérifier la condition :

(*) l'une des composantes irréductibles de X s'envoie surjectivement sur B.

Il est clair que cette condition doit être vérifiée "universellement", i.e. après tout changement de base $B' = \mathrm{Spec}\,R' \longrightarrow B$, où R' est le normalisé de R dans une extension finie K' de K : en effet, si X admet un point entier, il en est de même de $X' = X \times_B B'$. Ce raffinement n'est pas gratuit, comme le montre l'exemple

$$B = \mathrm{Spec}\,k[t],\ \mathrm{car}(k) \neq 2,\ X = \mathrm{Spec}\,k[x, 1/(x-1)],\ f(x) = x^2 - 1$$

(f est un revêtement double privé d'un point) : ici X est irréductible et f surjectif mais la condition (*) n'est plus vérifiée après le changement de base $g : B' \longrightarrow B$ où B' est le revêtement double "complet" $B' = \mathrm{Spec}\,k[x]$, $g(x) = x^2 - 1$.

En pratique, il suffit de tester la condition (*) après passage à une extension K' assez grande pour que les composantes de $X_{K^{alg}}$ soient définies sur K'. Remplaçant alors R par R' et X par une composante de X', on est amené finalement à imposer la

<u>Condition</u> (**) :

(i) f est surjectif;
(ii) X est irréductible;
(iii) X_K est géométriquement irréductible sur K.

3 - Le théorème de Rumely.

3.1. Nous allons maintenant chercher des conditions sur R pour que <u>tout</u> X vérifiant (**) admette un point entier. Soit d'abord L un fibré en droites sur B, et prenons pour X l'ouvert de L complémentaire de la section nulle. L'existence d'un point entier $Y \subset X$ implique alors que L est <u>d'ordre fini</u> dans $\mathrm{Pic}(B)$ (sa classe est annulée par le degré de Y sur B). Il convient donc de supposer que Pic(B) est un groupe <u>de torsion</u>, ainsi d'ailleurs que Pic (B') pour tout B-schéma B' fini et plat sur B, comme on le voit facilement en généralisant l'argument qui précède.

Cette condition est satisfaite par exemple dès que R est <u>local</u>; mais, même dans ce cas, on a le contre-exemple suivant, dû à Raynaud :

3.2. <u>Exemple</u>. Soient k un corps, R= Spec $k[t]_{(t)}$ l'anneau local à l'origine de $\mathbb{A}^1_k$, $g : E \longrightarrow B$ = Spec R une B-courbe elliptique non isotriviale, de section unité 0_E. Soit $A_0 \in E_0(k)$ un point rationnel de la fibre fermée E_0 de g. Considérons l'éclaté $\tilde{E}$ de A_0 dans E, et l'ouvert X de $\tilde{E}$ complémentaire du transformé strict de E_0. Il est clair que $f : X \longrightarrow B$ vérifie (**). Soit Y un point entier de X, de degré d>0 sur B : l'image Z de Y dans E est un diviseur dont l'intersection avec E_0 (vue comme diviseur sur E_0) est $d[A_0]$. Il existe d'autre part une unique section $P \in E(B)$ telle que $Z-d[0_E] \sim [P]-[0_E]$ (équivalence de diviseurs); spécialisant, on en déduit que $dA_0=P_0$ au sens de la loi de groupe de E_0. Or on sait (Lang-Néron) que le groupe E(B) est de type fini; par suite, si k est <u>non dénombrable</u> on peut choisir A_0 de manière que, pour tout d>0, $dA_0 \in E_0(k)$ ne soit pas dans l'image de $E_0(B)$. Ce qui précède montre qu'alors $X(R^{int})=\emptyset$.

3.3. Nous ferons désormais sur R les hypothèses suivantes :

<u>Condition</u> (***) :

(i) R est un anneau de Dedekind excellent;
(ii) pour tout R-schéma fini Z, le groupe Pic(Z) est <u>de torsion</u>;
(iii) pour tout idéal maximal P de R, le corps résiduel R/P est <u>extension algébrique d'un corps fini</u>.

<u>Théorème 1</u> (Rumely [3]). Sous les hypothèses (***), tout R-schéma séparé de type fini X vérifiant (**) possède un point entier.

3.4. On obtient le corollaire intéressant suivant en prenant pour X le schéma de modules sur $\mathbb{Z}$ des courbes lisses de genre donné g :

<u>Corollaire 1</u>. Pour tout $g \in \mathbb{N}$ il existe un corps de nombres K et une courbe C propre, lisse et géométriquement connexe de genre g sur K, ayant bonne réduction en chaque place de K.

3.5. <u>Remarques</u> :

1. La condition (***) est vérifiée notamment lorsque R est l'anneau des entiers d'un corps de nombres, ou l'anneau des fonctions d'une courbe affine lisse connexe sur un corps fini, ou un localisé d'un tel anneau.

2. Dans le corollaire ci-dessus, on peut imposer à C des conditions supplémentaires, par exemple (si $g \geq 3$) d'être non hyperelliptique modulo chaque place finie de K : il suffit pour cela de prendre pour X un ouvert convenable du schéma de modules.

3. Le théorème 1 s'étend, grâce au lemme de Chow, au cas où X n'est plus nécessairement un schéma mais seulement un <u>espace algébrique</u> (ou même un champ algébrique) sur B. Pour la même raison, il suffit d'établir le théorème lorsque X est un B-schéma <u>quasi-projectif</u>.

4. Cantor et Roquette [2] avaient établi ce théorème dans le cas particulier où X_K est une variété <u>unirationnelle</u>. La preuve de Rumely fait appel à sa "théorie des capacités" [4]; Roquette a, semble-t-il, une démonstration indépendante. La preuve géométrique esquissée ci-dessous est due à L. Szpiro et à l'auteur.

5. Dans [2] et [3], le théorème 1 est énoncé sous la forme "locale-globale" suivante. Pour tout idéal maximal P de R, soit R_P^{int} la fermeture intégrale du complété $\hat{R}_P$ dans une clôture algébrique de son corps des fractions; on suppose que R vérifie (***). Soit X un B-schéma séparé de type fini vérifiant (**)(iii) : alors, pour que $X(R^{int}) \neq \emptyset$ il faut et il suffit que $X(R_P^{int}) \neq \emptyset$ pour tout P.

   L'équivalence de ce résultat avec le théorème 1 est un exercice laissé au lecteur.

6. Le théorème 1 apporte une réponse positive au dixième problème de Hilbert sur $\mathbb{Z}^{int}$: il existe un algorithme permettant de décider si un système d'équations diophantiennes a une solution en entiers algébriques (cf. [3]).

4 - <u>Preuve du théorème de Rumely</u>.

4.1. Soit $f : X \longrightarrow B$ vérifiant (**); nous supposons X plongé dans un espace projectif $\mathbb{P}_B^N$ (cf. remarque 3 ci-dessus). Si X_K est de dimension ≥ 2, on commence par réduire sa dimension en le coupant par une hypersurface convenable (en prenant soin de préserver la condition (**); cf. [3] pour les détails). On suppose donc que $\dim(X_K)=1$ (autrement dit, $\dim(X)=2$).

4.2. Soit alors $\bar{X}$ l'adhérence de X dans $\mathbb{P}_B^N$, et posons $Z=\bar{X}-X$: c'est un fermé de $\bar{X}$ dont les composantes irréductibles sont de trois types :

- des points isolés;
- des composantes de dimension 1, <u>finies</u> sur B;
- des composantes <u>verticales</u>, i.e. de dimension 1 et contenues dans des fibres de $f : \mathcal{X} \longrightarrow B$.

4.3. Traitons d'abord le cas où Z n'a pas de composantes verticales. En d'autres termes, Z est fini sur B, et son groupe de Picard est donc de torsion (condition (***)(ii)). Si L désigne un faisceau inversible <u>ample</u> sur $\mathcal{X}$, on peut donc supposer, en remplaçant L par une puissance convenable, que

- la restriction de L à Z est triviale, donc admet une section s_0 partout non nulle;
- si I désigne l'Idéal de Z dans $\mathcal{X}$, on a $H^1(\mathcal{X}, I \otimes L)=0$, et par suite s_0 se relève en $s \in H^0(\mathcal{X},L)$.

Les zéros de s dans $\mathcal{X}$ forment un fermé F de X (puisque $F \cap Z$ est vide), propre (car fermé dans $\mathcal{X}$) et surjectif sur B (car L est ample); il est immédiat que l'une au moins des composantes irréductibles de F est un point entier.

4.4. Dans le cas général, on peut supposer $\mathcal{X}$ <u>régulier</u> (Abhyankar); on désigne alors par V la réunion des composantes verticales de Z, et l'on invoque le

<u>Théorème 2</u> (Artin [1], Raynaud). Il existe une contraction de V dans $\mathcal{X}$, i.e. un diagramme de B-schémas

$$\begin{array}{ccc} \mathcal{X} & \xrightarrow{\pi} & \mathcal{X}' \\ & {\scriptstyle f}\searrow \quad \swarrow{\scriptstyle f'} & \\ & B & \end{array}$$

avec les propriétés suivantes :

(i) f' est projectif;

(ii) $\pi(V)$ est un ensemble fini de points fermés de $\mathcal{X}'$;

(iii) $\pi_{|\mathcal{X}-V}$ est une immersion ouverte.

Il suffit donc de remplacer $\mathcal{X}$ par $\mathcal{X}'$ pour se ramener au cas précédent : ceci achève la démonstration du théorème 1.

4.5. <u>Remarques</u> :

1. Le théorème 2 vaut sous les hypothèses suivantes : B est un schéma régulier de dimension 1, $\mathcal{X}$ est régulier, f est projectif et plat à fibres géométriquement connexes et

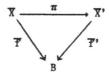

purement de dimension 1, V est un fermé de $\mathfrak{X}$, d'image dans B finie, ne contenant aucune fibre de $\mathfrak{f}$; enfin, ou bien B est le spectre d'un anneau de valuation discrète hensélien, ou bien les corps résiduels de B sont des extensions algébriques de corps finis.

2. Lorsque B est une courbe sur un corps fini, ce théorème est établi par Artin [1]. Le fait qu'il soit valable sous les hypothèses ci-dessus nous a été signalé par Raynaud.

3. Les hypothèses sur B sont essentielles : dans l'exemple 3.2, on ne peut pas contracter dans $\tilde{E}$ le transformé strict de E_0.

5 - Raffinements.

5.1. Sous les hypothèses du théorème 1, il est tentant d'imposer au point entier $Y \subset X$ des conditions locales, par exemple;

(i) (densité). Soit v une place de K; désignons par K_v^{alg} une clôture algébrique du complété K_v de K en v. Si U_v est un ouvert non vide de $X(K_v^{alg})$ (pour la topologie de v), peut-on trouver un point entier Y vérifiant $Y(K_v^{alg}) \subset U_v$?

(ii) Avec les mêmes notations, supposons que $X(K_v) \neq \emptyset$: peut-on choisir Y de manière que $Y \otimes_R K_v$ soit formé de points K_v-rationnels ?

Il faut naturellement imposer quelques restrictions. Dans (i), on doit supposer que U_v est invariant sous $Gal(K_v^{alg}/K_v)$; même ainsi, on méditera l'exemple :

$$R = \mathbf{Z},\ X = \operatorname{Spec} \mathbf{Z}[t, 1/t],\ K_v^{alg} = \mathbf{C},\ U_v = \{t \in \mathbf{C}^* \mid |t| < 1\}.$$

De même, la question (ii) n'est raisonnable que si X admet des points K_v-rationnels lisses. De fait, on montre :

Théorème 3. Soient R et X comme dans le théorème 1. Soit Σ un ensemble fini de places de K (archimédiennes ou non), disjoint de l'ensemble $Max(R)$ des idéaux maximaux de R. Pour tout $v \in \Sigma$, on se donne :

- une extension finie galoisienne L_v de K_v;
- un ouvert non vide U_v de $X(L_v)$, formé de points lisses et invariant sous $Gal(L_v/K_v)$.

On suppose qu'il existe une place de K n'appartenant ni à Σ, ni à Max(R).

Alors X contient un point entier Y tel que, pour tout $v\in\Sigma$, $Y \otimes_R L_v$ soit formé de points L_v-rationnels, tous dans U_v.

5.2. Ainsi, si $P\in\mathbb{Z}[X_1,\dots,X_n]$ est un polynôme <u>primitif</u> (1.1), p un nombre premier, I un intervalle ouvert non vide de $\mathbb{R}$, il existe :

- des entiers algébriques $x_1,\dots,x_n$ tels que $P(x_1,\dots,x_n)$ soit inversible dans $\mathbb{Z}^{int}$ et tels que l'extension de $\mathbb{Q}$ engendrée par les x_i soit <u>décomposée</u> en p;

- des nombres $y_1,\dots,y_n$ entiers sur $\mathbb{Z}[1/p]$, tels que $P(y_1,\dots,y_n)$ soit inversible dans $\mathbb{Z}^{int}[1/p]$, et que tous les conjugués des y_i soient dans I.

Le théorème 3 est prouvé dans [2] lorsque K est un corps de nombres, sous l'hypothèse "X_K unirationnelle", avec $L_v=K_v$; on trouvera dans [3] l'énoncé obtenu en remplaçant L_v par K_v^{alg}. La cas général est en préparation.

(*) p. : 147. Unité de recherche associée au C.N.R.S. n° 752

BIBLIOGRAPHIE

[1] M. Artin.- Some numerical criteria for contractibility of curves on algebraic surfaces, Amer. J. Math. 84 (1962), 485-496.

[2] D. Cantor et P. Roquette.- On Diophantine Equations over the Ring of All Algebraic Integers, J. of Number Theory 18 (1984), 1-26.

[3] R. Rumely.- Arithmetic over the ring of all algebraic integers, J. reine und angew. Math. 368 (1986), 127-133.

[4] R. Rumely.- Capacity Theory on Algebraic Curves, Lecture Notes in Math. (Springer), à paraître.

[5] T. Skolem.- Lösung gewisser Gleichungen in ganzen algebraischen Zahlen, insbesondere in Einheiten, Skrifter Norske Videnskaps-Akademii Oslo, Mat. Naturv. Kl. 10 (1934).

Laurent MORET-BAILLY
Département de Mathématique
Université Paris-Sud
91405 ORSAY CEDEX

Approximants de Padé et mesures effectives d'irrationalité

Georges RHIN

1 - Introduction.

Les approximants de Padé des fonctions hypergéométriques ont été utilisés pour l'étude en des points rationnels $z=p/q$ des approximations diophantiennes des valeurs de ces fonctions. Cette méthode puissante a donné des résultats rappelés au paragraphe 3. Nous montrerons au paragraphe 4 comment les approximants de Padé peuvent être remplacés par d'autres polynômes qui fournissent de meilleures mesures d'irrationalité, améliorant certains résultats de G.V. Chudnovsky. Les définitions nécessaires sont rappelées au paragraphe 2.

2 - Définitions.

Soit α un réel non rationnel. On dit que $\mu=\mu(\alpha)$ est une mesure effective d'irrationalité de α si pour tout $\epsilon>0$ il existe $q_0(\epsilon)>0$ effectivement calculable tel que

$$\forall (p,q)\in\mathbb{Z}^2 \quad q>q_0(\epsilon) \implies |\alpha-p/q|>q^{-\mu-\epsilon}.$$

Les approximants de Padé sont de deux types.

Soient $f_1,\dots,f_n$ des fonctions analytiques en 0 (ou des séries formelles).

Approximants de Padé de type I.
Un système d'approximants de Padé de type I de $(f_1,\dots,f_n)$ d'ordre N est un n-uplet $(A_1,\dots,A_n)$ de polynômes non tous nuls vérifiant

(1) $$\deg A_i \leq N$$

(2) $$1 + \mathrm{ord}_0(\Sigma A_i f_i) \geq n(N+1).$$

La fonction $R=\Sigma A_i f_i$ s'appelle le reste de l'approximation.

Approximants de Padé de type II.
Un système d'approximants de Padé de type II de $(f_1,\dots,f_n)$ d'ordre N est un n-uplet $(B_1,\dots,B_n)$ de polynômes non tous nuls vérifiant

$$(3) \qquad \deg B_i \leq (n-1)N$$

$$(4) \qquad \forall\, k,\ell \ \ \mathrm{ord}_0\,(B_\ell f_k - B_k f_\ell) \geq n\,N+1$$

C'est le calcul explicite d'un système d'approximants de Padé de fonctions exponentielles qui a permis à Hermite de démontrer en 1873 la transcendance de e.

3 - Approximants de Padé et fonctions hypergéométriques.

Euler et Gauss ont donné des développements en fraction continue de quotients de deux fonctions hypergéométriques et en 1909 Padé construisit les approximants de Padé de la fonction hypergéométrique :

$${}_2F_1(1,b;c;x) = \sum_{n\geq 0} \frac{(b)_n}{(c)_n}\, x^n$$

où $c>0$ et $(b)_n = b(b+1)\dots(b+n-1)$.

Nous considérons les cas particuliers suivants :

$$\log(1-z) = z\ {}_2F_1\ (1,1;2;z)$$

$$\log\left(\frac{1+z}{1-z}\right) = 2z\ {}_2F_1\ \left(1,\tfrac{1}{2};\tfrac{3}{2};z^2\right)$$

$$\mathrm{Arc\ tg}(z) = z\ {}_2F_1\ \left(1,\tfrac{1}{2};\tfrac{3}{2};-z^2\right)$$

et

$$(1-z)^\omega = {}_2F_1\ (1,-\omega;1;z).$$

En 1953 [14] Mahler reprenait les calculs d'Hermite qui par un changement de variable donnent un système d'approximants de Padé des fonctions $1, \log(1+z),\dots,\log(1+z)^{m-1}$. Ceci lui permit de donner des mesures d'irrationalité de logarithmes de nombres rationnels ou de nombres d'un corps quadratique imaginaire. Il obtint alors $\mu(\pi)=42$. En 1973, améliorant l'estimation du reste, Mignotte [15] démontra que $\mu(\pi)=20$. Par la suite le reste sera exprimé sous forme d'intégrale de Laplace par Chudnovsky et E. Reyssat qui démontra [16] :

Théorème 1. Soient $\alpha=p/q$ un rationnel >1, ϵ un réel >0 et d, H, $m\geq 2$ trois entiers. Alors pour tout nombre algébrique ξ de degré d et de hauteur $\leq H$ on a $|\log(\alpha)-\xi| > H^{-c-\epsilon}$ où

$$c = \frac{m(m-1)(\log(\alpha^{1/m}-1)-\log(\alpha^{1/m}+1))}{d(m-1+\log b)+(d-1)m\ \log(\alpha^{1/m}+1)+\log(\alpha^{1/m}-1)}$$

pourvu que c soit positif (ce qui est le cas si m est assez grand) et que H soit assez grand en fonction de r, m, ϵ et d (ces conditions peuvent être explicitées).

Ce résultat permet par exemple d'obtenir $\mu(\log 3)= 14,7$. En 1964 Baker démontre [4] $\mu(\log 2)= 12,5$. Ce résultat est amélioré en 1978 par Apéry [2,11] qui donne $\mu(\log 2)= 4,622\ldots$ et indépendamment par Danilov [12]. On trouve les calculs effectifs correspondants dans un article d'Alladi et Robinson de 1980 [1]. Ils considèrent l'intégrale

$$I_n = z^{2n+1} \int_0^1 \frac{t^n(1-t)^n}{(1-zt)^{n+1}}\, dt = P_n(z)\log(1-z) - Q_n(z)$$

qui fournit les approximants de Padé de $\log(1-z)$. P_n et Q_n sont des polynômes à coefficients rationnels dont le dénominateur commun divise ppcm $(1,2,\ldots,n)$. Lorsque z est rationnel $Q_n(z)/P_n(z)$ fournit une approximation rationnelle de $\log(1-z)$. On obtient alors :

<u>Théorème 2</u>. Soit K le corps des rationnels ou un corps quadratique imaginaire et R l'anneau des entiers de K. Soient t et s deux éléments non nuls de R.

On pose
$$\gamma_1 = \min \left\{\left|\frac{(1\pm\sqrt{1-t/s})^2}{t/s}\right|\right\}$$

et
$$\gamma_2 = \max \left\{\left|\frac{(1\pm\sqrt{1-t/s})^2}{t/s}\right|\right\}.$$

Alors si t et s vérifient les deux conditions suivantes :

$$\frac{t}{s} \notin [1,\infty) \tag{5}$$

$$\gamma_1\, |t|\, e < 1 \tag{6}$$

le nombre $\log(1-t/s)$ n'appartient pas à K et

$$\mu = 1 + \frac{\log(\gamma_2|t|)+1}{\log(\gamma_2/|t|)-1}$$

est une mesure effective d'irrationalité de $\log(1-t/s)$.

On obtient comme corollaire $\mu(\pi/\sqrt{3})= 8,309\ 986\ldots$ En 1980 G.V. Chudnovsky donne une théorie générale utilisant les groupes de monodromie des équations différentielles satisfaites par les fonctions hypergéométriques [7] et démontre qu'il est possible d'obtenir les approximants de Padé de type I de systèmes plus généraux de fonctions hypergéométriques. En 1982, Chudnovsky donne $\mu(\log 2)= 4,269\ 6549$ [8] et $\mu(\pi/\sqrt{3})= 5,817\ldots$ [9] et annonce $\mu(\log 2)= 4,134\ 400\ 029$ et $\mu(\pi/\sqrt{3})= 5,792\ldots$. Il explique que ces résultats s'obtiennent aussi par des considérations de monodromie mais que les polynômes obtenus ne sont plus des approximants de Padé des fonctions ypergéométriques considérées. D'autre part les calculs explicites ne sont pas aisés car les approximations utilisées vérifient des relations de récurrence linéaires compliquées.

Dans le cas de plusieurs logarithmes G. Rhin et P. Toffin ont donné [17] les approximants de Padé de type II de 1, $\log(1+\alpha_1 z),\ldots,\log(1+\alpha_1 z)$. On obtient alors le résultat d'approximation suivant :

<u>Théorème 3</u>. Soient $\alpha_1,\ldots,\alpha_r$ r nombres rationnels non nuls, Q un entier positif tel que pour $1\leq i\leq r$, $Q\alpha_i\in\mathbb{Z}$ et $\alpha = \max_{1\leq i\leq r} |\alpha_i| \leq 1/(r+2)$. Il existe deux constantes positives θ, γ ne dépendant que de r et de α telles que pour tout $\epsilon>0$

$$\text{si} \quad \omega = \theta\,\alpha^{r+1}\,Q^r\,e^{r(1+\epsilon)} < 1$$

on ait $|u_0 + u_1\log(1+\alpha_1)+\ldots+u_r\log(1+\alpha_r)| \geq c\,H^{-\mu}$ où $u_0,u_1,\ldots,u_r$ sont entiers, $H = \max_{1\leq i\leq r} |u_i| > 0$,

$$\mu = -\frac{n(1+\epsilon+\log Q)+\log\gamma}{\log\omega} \quad \text{et} \quad c = c(\alpha,r,\epsilon).$$

On obtient par exemple que

$$|u_0+u_1\log(3/4)+u_2\log(5/4)| \geq H^{-88}e^{-196}$$

alors que la méthode de baker qui donne des minorations des formes linéaires de logarithmes de nombres algébriques [5] donne dans ce cas particulier une moins bonne estimation (on trouve 3.10^{21} au lieu de 88 [19]).

Dans le cas des racines cubiques d'entiers Baker a obtenu en 1964 [3] $\mu(\sqrt[3]{2})= 2,947\ldots$ en utilisant des approximations de Padé de $(1-z)^{1/3}$. En 1979 Chudnovsky [10] a démontré en étudiant les facteurs premiers communs aux numérateurs des approximations rationnelles de $\sqrt[3]{2}$ donnés par les approximants de Padé que $\mu(\sqrt[3]{2})= 2,429\ldots$. Rappelons que le théorème de Roth donne une mesure d'irrationalité de tous les nombres algébriques égale à 2 si l'on n'exisge pas que $q_0(\epsilon)$ soit effectivement calculable.

4 - <u>Généralisation de la méthode d'Alladi-Robinson</u>.

Nous allons, dans l'étude de l'irrationalité de $\log(1-z)(z\in\mathbb{Q}^*, z\notin[1,\infty))$, remplacer, dans l'intégrale I_n donnée au paragraphe 3, le polynôme $t^n(1-t)^n$ par un polynôme plus général satisfaisant certaines conditions arithmétiques.

Soient d un entier positif tel que dz soit entier et $\Delta=\mathrm{ppcm}(d,d-dz)$. Soit H_n un polynôme de $\mathbb{Z}[x]$, de degré inférieur ou égal à $2n$, appartenant à l'idéal

$$((x,\Delta)\mathbb{Z}[x])^n$$

et donc de la forme

$$H_n(x)=\sum_{j=0}^{n} b_j\,\Delta^{n-j}x^j+\sum_{j=n+1}^{2n} b_j\,x^j$$

où les b_j sont des entiers pour $0\leq j\leq 2n$.

On étudie l'intégrale

$$I_n=\int_0^1 \frac{H_n(d-dzt)}{d^n(1-zt)^{n+1}}\,dt$$

et

$$-z\,I_n=b_n\log(1-z)+\sum_{j=n+1}^{2n}\frac{bj}{j-n}\left((d-dz)^{j-n}-d^{j-n}\right)-\sum_{j=0}^{n-1}\frac{bj}{n-j}\left(\left(\frac{\Delta}{d-dz}\right)^{n-j}-\left(\frac{\Delta}{d}\right)^{n-j}\right).$$

Alors si $d_n=\mathrm{ppcm}(1,2,\ldots,n)$, $-z\,d_nI_n\in\mathbb{Z}[\log(1-z)]$. Sachant que $b_n=-\frac{z}{2\pi i}\int_{|u+\frac{1}{z}|=p}\frac{H_n(d-dzu)}{d^n(1-zu)^{n+1}}\,du$, si le polynôme H_n est assez simple on obtient des majorations du type

$$0<|I_n|\leq e^{an},\quad |b_n|\leq e^{bn}.$$

Il suffit alors d'utiliser des versions explicites du théorème des nombres premiers pour majorer d_n et l'on obtient si $a<-1$.

$$\mu(\log(1-z))=-\frac{b+1}{a+1}+1.$$

Pour $\log 2$, $z=-1$, $d-dz\;t=1+t$ et $\Delta=2$. Si $H_n(t+1)=t^n(1-t)^n$ on obtient le résultat $\mu(\log 2)=4{,}622\ldots$ donné par Apéry etc... . Si $H_n(t+1)=[t(1-t)]^{[0{,}88n]}(t^2+2t-1)^{2[0{,}06n]}$ on obtient le premier résultat de Chudnovsky $\mu(\log 2)=4{,}265\;174\;29\ldots$ et si

$$H_n(t+1)=[t(1-t)]^{[0{,}9n]}(6t^2-5t+1)^{2[0{,}05n]}$$

on obtient $\mu(\log 2)= 4,134\,400\,029\ldots$ qui est probablement le deuxième résultat de Chudnovsky. En utilisant un polynôme H_n plus compliqué (donné en appendice) on peut démontrer que $\mu(\log 2)= 4,0765$. Il semble raisonnable d'espérer que cette méthode permettra de montrer que $\mu(\log 2)<4$.

La méthode ci-dessus se généralise à l'étude de l'indépendance linéaire de plusieurs logarithmes. Soient $\alpha_1,\alpha_2,\ldots,\alpha_r$ r rationnels non nuls tels que 1, $\log(1-\alpha_1),\ldots,\log(1-\alpha_r)$ soient $\mathbb{Q}$-linéairement indépendants. Soit d un entier positif tel que $d\alpha_i$ soit entier pour tout i et $\Delta= \underset{1\le i\le r}{\mathrm{ppcm}}\,(d\alpha_i,d-d\alpha_i)$.

Soit H_n un polynôme qui vérifie les conditions du début du paragraphe. Alors les intégrales

$$I_n(\alpha_i)= \int_0^1 \frac{H_n(d-d\alpha_i t)}{d^n(1-\alpha_i t)^{n+1}}\,dt$$

sont telles que $d_n\alpha_i I_n(\alpha_i)= P_{n,i} - d_n b_n \log(1-\alpha_i)$ où $p_{n,i}$ est entier. On remarque que dans le cas des approximants de Padé le polynôme du numérateur est de degré $(r+1)n$ et que le dénominateur des $I_n(\alpha_i)$ divise d_{rn} (au lieu de d_n). En étudiant $\log(2/3)$ et $\log(4/3)$ on obtient :

<u>Proposition</u>. Soient u_0, u_1, u_2 trois entiers tels que $H= \max(|u_1|,|u_2|) \ge 2$. Alors la forme $\Lambda= u_0 + u_1 \log 2 + u_2 \log 3$ vérifie

$$|\Lambda| \ge H^{-13,3}. \tag{7}$$

De plus pour $H \ge H_0$ (H_0 effectivement calculable)

$$|\Lambda| \ge H^{-7,616}. \tag{8}$$

La démonstration de (7) utilise les meilleures approximations simultanées $q\log 2-p_1$ et $q\log 3-p_2$ de $\log 2$ et $\log 3$ pour $q\le 7,32\ 10^{12}$. Ces calculs ont été effectués par E. Dubois et P. Toffin avec Macsyma en utilisant un algorithme de E. Dubois. La méthode de Baker donne ici 2^{50} au lieu de $13,3$. Le résultat (8) améliore en particulier la mesure d'irrationalité de $\log 3$ donnée par E. Reyssat.

La méthode s'applique aussi aux nombres

$$\int_0^1 \frac{t^{-1/2}}{1-zt}\,dt,\quad z\in\mathbb{Q}^*,\quad z\notin[1,\infty).$$

On étudie alors l'intégrale

$$I_n = \int_0^1 \frac{t^{-1/2} H_n(d-dzt)}{d^n(1-zt)^{n+1}} dt$$

où H_n est à coefficients entiers de degré inférieur ou égal à 1,5n et dans l'idéal $((x,\Delta)\mathbb{Z}[x])^n$ avec Δ= ppcm(4d,d-dz).

Pour z=-1/3, Δ=12. Si l'on prend

$$H_n(3+t) = (t(1-t))^{[0,75n]}.12^{[0,25]+2}$$

on obtient le résultat de Chudnovsky $\mu(\pi/\sqrt{3})$= 5,817. Si l'on ne fait intervenir dans les facteurs de H_n que des polynômes qui ont toutes leurs racines dans l'intervalle [0,1] on peut utiliser une méthode d'optimisation introduite par C.H. Smyth [18] qui permet une recherche quasi-automatique des "bons facteurs" de H_n. On obtient alors $\mu(\pi/\sqrt{3})$= 4,97. On peut aussi obtenir avec la même méthode $\mu(\pi)$=23,918 (avec z=-1 et Δ=4) ce qui est moins bon que le résultat de Mignotte.

Pour z=1/3, Δ=2 le polynôme

$$H_n(3-t) = 144(2t(1-t))^{[0,5n]}.(5t-3)^{[0,5n]}$$

donne $\mu(\sqrt{3}\log(2+\sqrt{3}))$= 17,207... , alors que la méthode des approximants de Padé utilisée par M. Huttner dans ces cas [13] ne permet pas de donner de mesure d'irrationalité de $\sqrt{3}\log(2+\sqrt{3})$.

<u>Appendice</u> :

1. Le polynôme H_n utilisé pour montrer que $\mu(\log 2)$= 4,0765 est donné par la formule

$$H_n(1+t) = \prod_{i=1}^{6} 2\, P_i(t)^{[a_i,n]}$$

où

$P_1 = t(1-t)$	$a_1 = 0{,}84943$
$P_2 = t^2+2t-1$	$a_2 = 0{,}02401$
$P_3 = 6t^2-5t+1$	$a_3 = 0{,}091$
$P_4 = 7t^2-6t+1$	$a_4 = 0{,}02068$
$P_5 = 13t^2-11t+2$	$a_5 = 0{,}0113$
$P_6 = 2t^4+3t^3+3t^2-5t+1$	$a_6 = 0{,}00179$

2. Pour démontrer (8) on étudie les deux intégrales

$$\int_2^3 \frac{H_n(x)}{x^{n+1}}\,dx \quad \text{et} \quad \int_3^4 \frac{H_n(x)}{x^{n+1}}\,dx$$

où $H_n(x) = 12^7 \prod_{i=1}^{6} Q_i(x)^{[b_i n]}$ avec

$Q_1 = x-3$	$b_1 = 0,704324$
$Q_2 = x-2$	$b_2 = 0,552418$
$Q_3 = x-4$	$b_3 = 0,447582$
$Q_4 = 5x-12$	$b_4 = 0,109072$
$Q_5 = 17x^2-102x+144$	$b_5 = 0,038934$
$Q_6 = 19x^2-108x+144$	$b_6 = 0,054\,368$

N.B. : Les calculs peuvent être vérifiés facilement sur un micro ordinateur.

BIBLIOGRAPHIE

[1] K. Alladi et M.L. Robinson.- Legendre polynomials and irrationality. J. Reine Angew. Math. 318 (1980), 137-155.

[2] R. Apéry.- Irrationalité de $\zeta(2)$ et $\zeta(3)$. Journées arithmétiques de Luminy, Astérisque 61 1979, 11-13.

[3] A. Baker.- Rational approximation to certain algebraic numbers. Proc. Lond. Math. Soc. 14 (1964), 385-393.
et
A. Baker.- Rational approximations to $\sqrt[3]{2}$ and other algebraic numbers. Quart. J. Math. 15 (1964), 376-383.

[4] A. Baker.- Approximation to the logarithm of certain rational numbers. Acta Arith. 10 (1964), 315-323.

[5] A. Baker.- Transcendental Number Theory. 2ème éd. Cambridge Univ. Press. 1979, voir aussi [19].

[6] F. Beukers.- Legendre polynomials in irrationality proofs. Bull. Austral. Math. Soc. 22 (1980), 431-438.

[7] G.V. Chudnovsky.- Padé approximation and the Riemann monodromy problem in Bifurcation Phenomena in Mathematical Physics and Related Topics. (1980) Reidel, 449-510

[8] D.V. Chudnovsky et G.V. Chudnovsky.- Padé and Rational Approximation to systems of Functions and their Arithmetic Applications. Lect. Notes in Math. (1982), n° 1052, 37-84.

[9] G.V. Chudnovsky.- Number Theoretic Applications of Polynomials with Rational Coefficients Defined by Extremaly Conditions. Arithmetic and Geometry. Birkhäuser (1983), 61-105.

[10] G.V. Chudnovsky.- On the method of Thue Siegel. Annals of Math. 117 (1983), 325-382.

[11] H. Cohen.- Démonstration de l'irrationalité de $\zeta(3)$ (d'après R. Apéry). Séminaire de Théorie des Nombres. Grenoble 1978, 9 pages.

[12] V. Danilov.- Rational approximations of some functions at rational points. Math. Zametki 24 (1978) N° 4 et Math. Notes USSR 24 (1979), 741-744.

[13] M. Huttner.- Problème de Riemann et irrationalité d'un quotient de deux fonctions hypergéométriques de Gauss. C.R. Acad. Sci. Paris 302 I, 603-606.

[14] K. Mahler.- On the approximation of logarithms of algebraic numbers. Phil. Trans. Roy. Soc. London A 245 (1953), 371-398.

[15] M. Mignotte.- Approximations rationnelles de π et quelques autres nombres. Journées arithmétiques de Grenoble (1973), Bull. Soc. Math. France Mémoire 37 (1974), 121-132.

[16] E. Reyssat.- Mesures de transcendance pour les logarithmes de nombres rationnels in Approximations Diophantiennes et Nombres Transcendants. Progress in Math. Birkhäuser (1983), 235-245.

[17] G. Rhin et P. Toffin.- Approximants de Padé simultanés de logarithmes. Journal of Number Theory 24, (1986), 284-297.

[18] C.J. Smyth.- On the Measure of Totally Real Algebraic Integers II. Math. of Comp. 37 (1981), 205-208.

[19] M. Waldschmidt.- A lower bound for linear forms in logarithms. Acta Arithmetica 37 (1980), 257-283.

Georges RHIN
Département de Mathématiques
Université de Metz
Faculté des Sciences
Ile du Saulcy
57045 METZ 1

DESCENTS ON ELLIPTIC CURVES WITH COMPLEX MULTIPLICATION

Karl RUBIN*

To obtain information on the group of rational points or the Tate-Safarevic group of an elliptic curve E defined over $\mathbb{Q}$, one usually does a "descent", at some prime(s) p, to determine the Selmer group of E relative to p (or powers of p). There is a finite algorithm for carrying out such a descent, but with few exceptions (see [2] for examples with p=2 and [10], [12] for some with p=3) this algorithm is completely impractical in actual examples.

In [5] Coates and Wiles introduced new methods which provide a much more practical descent procedure for elliptic curves with complex multiplication, for odd primes p of good, ordinary reduction which are not anomalous for E (p is anomalous if p divides $|\tilde{E}(\mathbb{F}_p)|$, where $\tilde{E}$ denotes the reduction of E modulo p). See [1] for some numerical examples. In this note we show how to use an explicit reciprocity law to extend the Coates-Wiles methods to supersingular and anomalous primes as well (still only for elliptic curves with complex multiplication). As numerical examples (see § 2) we show that for the following curves E and primes p, the p-part of the Tate-Safarevic group of E over $\mathbb{Q}$ is trivial :

E	CM *field*	p	*type*
$y^2 = x^3 - x$	$\mathbb{Q}(\sqrt{-1})$	3	supersingular
$y^2 + y = x^3$	$\mathbb{Q}(\sqrt{-3})$	5	supersingular
$y^2 + xy = x^3 - x^2 - 2x - 1$	$\mathbb{Q}(\sqrt{-7})$	3,5	supersingular
$y^2 = 4x^3 + 4x^2 - 28x - 41$	$\mathbb{Q}(\sqrt{-11})$	3	ordinary, anomalous

§1. Let E be an elliptic curve defined over $\mathbb{Q}$ with complex multiplication by the integers $\mathcal{O}$ of an imaginary quadratic field K. For any nonzero rational integer m the exact sequence

$$0 \longrightarrow E_m \longrightarrow E(\bar{\mathbb{Q}}) \xrightarrow{m} E(\bar{\mathbb{Q}}) \longrightarrow 0$$

gives rise to a $G_{\mathbb{Q}} = \mathrm{Gal}(\bar{\mathbb{Q}}/\mathbb{Q})$-cohomology exact sequence

$$(1)\qquad 0 \longrightarrow E(\mathbb{Q})/mE(\mathbb{Q}) \longrightarrow H^1(G_{\mathbb{Q}},E_m) \longrightarrow H^1(G_{\mathbb{Q}},E(\overline{\mathbb{Q}}))_m \longrightarrow 0$$

where $H^1(G_{\mathbb{Q}},E(\overline{\mathbb{Q}}))_m$ denotes the m-torsion subgroup of $H^1(G_{\mathbb{Q}},E(\overline{\mathbb{Q}}))$. Define the Tate-Safarevic group

$$Ш_m(\mathbb{Q}) = \ker\left[H^1(G_{\mathbb{Q}},E(\overline{\mathbb{Q}}))_m \longrightarrow \bigoplus_{\ell} H^1(G_{\mathbb{Q}_\ell},E(\overline{\mathbb{Q}}_\ell))\right]$$

and define the Selmer group relative to m, $S_m(\mathbb{Q})$, to be the inverse image of $Ш_m(\mathbb{Q})$ in $H^1(G_{\mathbb{Q}},E_m)$ in (1). Thus

$$(2)\qquad 0 \longrightarrow E(\mathbb{Q})/mE(\mathbb{Q}) \longrightarrow S_m(\mathbb{Q}) \longrightarrow Ш_m(\mathbb{Q}) \longrightarrow 0$$

In a completely analogous way we can define $Ш_\alpha(K)$ and $S_\alpha(K)$ for any nonzero integer α of K.

Fix a rational prime $p\geq 3$ where E has good reduction and a prime $\mathfrak{p}$ of K dividing p. Let $\Omega\in\mathbb{R}^*$ be the fundamental real period of a minimal model of E, and $L(E,s)$ the Hasse-Weil L-function of E over $\mathbb{Q}$. Then $\Omega^{-1}L(E,1)$ is known to be a rational number, integral at p. Our main descent criterion is the following theorem, which will be proved in §3.

<u>Theorem 1</u>. Let $F= K(E_p)$, and let A denote the ideal class group of F and $r = \dim_{\mathcal{O}/\mathfrak{p}} \mathrm{Hom}(A,E_{\mathfrak{p}})^{\mathrm{Gal}(F/K)}$. Then

$$r \leq \dim_{\mathbb{F}_p} S_p(\mathbb{Q}) \leq r+1.$$

If $\Omega^{-1}L(E,1) \not\equiv 0$ (modulo p) then $S_p(\mathbb{Q}) \cong \mathbb{F}_p^r$.

<u>Remarks</u> :

1) When p is ordinary and not anomalous for E this theorem is essentially contained in [5]. Compare also Theorem II.19 of [1].

2) Cassels has shown [3] that if $Ш_{p^\infty}(\mathbb{Q}) = \varinjlim Ш_{p^n}(\mathbb{Q})$ is finite then $\dim_{\mathbb{F}_p} Ш_p(\mathbb{Q})$ is even. Thus if r=0 the first part of Theorem 1 implies that either $Ш_p(\mathbb{Q})=0$ or $Ш_{p^\infty}(\mathbb{Q})$ is infinite. Using the techniques of this paper it is possible to show that if r=0 and $L(E,1)\neq 0$ then $Ш_{p^\infty}(\mathbb{Q})$ is finite. In this case $E(\mathbb{Q})$ is also finite by the Coates-Wiles theorem [5]. Thus if r=0 the condition $\Omega^{-1}L(E,1)\not\equiv 0$ (mod p) in Theorem 1 could be weakened to $L(E,1)\neq 0$, but (if one believes the Birch and Swinnerton-Dyer conjecture) this is no

more helpful in applications. Of course, the Birch and Swinnerton-Dyer conjecture predicts that the condition $\Omega^{-1}L(E,1)\not\equiv 0 \pmod p$ alone is equivalent to $S_p(\mathbb{Q})=0$.

Let h denote the class number of $F=K(E_{\mathfrak{p}})$.

Corollary 2. If $\mathbb{N}\mathfrak{p}\nmid h$ and $\Omega^{-1}L(E,1)\not\equiv 0$ (modulo p) then $Ш_p(\mathbb{Q})=0$.
Proof. Since $\mathfrak{p}$ is a prime of good reduction, $Gal(F/K)$ acts transitively on $E_{\mathfrak{p}}-0$ (see Lemma 5 of [5]). Therefore if $f\in Hom(A,E_{\mathfrak{p}})^{Gal(F/K)}$ is nonzero, f is surjective, so $|E_{\mathfrak{p}}|=\mathbb{N}\mathfrak{p}$ must divide h. Thus $r=0$ and Theorem 1 applies.

Corollary 3. If $rank_{\mathbb{Z}}\,E(\mathbb{Q})\geq r+1$ (for example, if $\mathbb{N}\mathfrak{p}\nmid h$ and $E(\mathbb{Q})$ is infinite) then $rank_{\mathbb{Z}}\,E(\mathbb{Q})=r+1$ and $Ш_p(\mathbb{Q})=0$.
Proof. This follows immediately from Theorem 1 and (2).

§2. In this section we use Corollary 2 and analytic class number bounds to prove the five examples given in the introduction. We have the four elliptic curves, with their conductors (N) and their identifying numbers from [14] (#) :

E	N	#
$E^{(1)}: y^2= x^3-x$	32	32A
$E^{(2)}: y^2+y= x^3$	27	27A
$E^{(3)}: y^2+xy= x^3-x^2-2x-1$	49	49A
$E^{(4)}: y^2= 4x^3+4x^2-28x-41$	1936	twist of 121D by -1

and the following table of data :

E	K	f	$\Omega^{-1}L(E,1)$	p	$\mathfrak{p}$	$D_{F/\mathbb{Q}}$
$E^{(1)}$	$\mathbb{Q}(\sqrt{-1})$	$(1+i)^3$	1/4	3	(3)	$2^{32}3^{14}$
$E^{(2)}$	$\mathbb{Q}(\sqrt{-3})$	(3)	1/3	5	(5)	$3^{60}5^{46}$
$E^{(3)}$	$\mathbb{Q}(\sqrt{-7})$	$(\sqrt{-7})$	1/2	3	(3)	$3^{14}7^{12}$
"	"	"	"	5	(5)	$5^{46}7^{36}$
$E^{(4)}$	$\mathbb{Q}(\sqrt{-11})$	$(4\sqrt{-11})$	1	3	$((1-\sqrt{-11})/2)$	$2^4 3.11^3$

Here f denotes the conductor of the Hecke character of K attached to E; it is related to N by a formula of Shimura [11] : $N=|D_{K/\mathbb{Q}}|\mathbb{N}_{K/\mathbb{Q}}(f)$. The values $L(E,1)$ and Ω are easily computed. It remains to explain the last column, the absolute discriminant of $F=K(E_{\mathfrak{p}})$. It is easy to see that F/K is a cyclic extension and the character group of $Gal(F/K)$ is generated by the character χ giving the action of $Gal(F/K)$ on $E_{\mathfrak{p}}$. The arguments of [5] (especially Lemma 4) show that χ has order $\mathbb{N}\mathfrak{p}-1$ and conductor $\mathfrak{f}\mathfrak{p}$; further, if $1 \leq i \leq \mathbb{N}\mathfrak{p}-2$, χ^i has conductor given as follows :

$E^{(1)}$		$E^{(2)}$		$E^{(3)}, E^{(4)}$	
i(mod 4)	cond(χ^i)	i(mod 6)	cond(χ^i)	i(mod 2)	cond(χ^i)
1,3	$\mathfrak{f}\mathfrak{p}$	1,2,4,5	$\mathfrak{f}\mathfrak{p}$	1	$\mathfrak{f}\mathfrak{p}$
2	$2\mathfrak{p}$	3	$\sqrt{-3}\mathfrak{p}$	0	$\mathfrak{p}$
0	$\mathfrak{p}$	0	$\mathfrak{p}$		

Using these values one can compute the discriminant $D_{F/\mathbb{Q}}$ by the formulae $D_{F/K} = \prod_{i=1}^{\mathbb{N}\mathfrak{p}-1} \text{cond}(\chi^i)$, $D_{F/\mathbb{Q}} = D_{K/\mathbb{Q}}^{[F:K]} \mathbb{N}_{K/\mathbb{Q}}(D_{F/K})$, giving the numbers in the table above.

We now use Odlyzko's discriminant bound ([7], Theorem 1) which for the Hilbert class field of F gives

$$(D_{F/\mathbb{Q}})^h \geq (21.8)^{2(\mathbb{N}\mathfrak{p}-1)h} e^{-70}.$$

Solving for h yields the following upper bounds :

	$E^{(1)}$	$E^{(2)}$	$E^{(3)}$	$E^{(3)}$	$E^{(4)}$
$\mathbb{N}\mathfrak{p}=$	9	25	9	25	3
$h <$	5.96	8.78	6.62	18.21	55.50

Therefore in our first four examples Corollary 2 applies to show that $Ш_p(\mathbb{Q})=0$.

We need a different method to show in the last example that $3|h$. Write $\pi = (1-\sqrt{-11})/2$, so π is a generator of the ideal $\mathfrak{p}$. The nonzero elements of $E_{\mathfrak{p}}^{(4)}$ are $((-2+\sqrt{-11})/\pi, \pm(-\sqrt{-11}/\pi)^{3/2})$ (these are points of order 3, not integral at $\mathfrak{p}$), so we have $F= K(\sqrt{\alpha})$ where $\alpha= -\pi\sqrt{-11}= (-11-\sqrt{-11})/2$.

By Minkowski's theorem, any ideal class contains an integral ideal of norm less than $(4/3.14)^2(4!/4^4)\sqrt{D_{F/\mathbb{Q}}} < 38.42$. Using the description $F= K(\sqrt{\alpha})$ it is an exercise to find the primes of F of norm less than 39, namely (using * to denote conjugation by $Gal(F/K)$) :

prime	$\mathbb{N}_{F/K}$	$\mathbb{N}_{F/K}$	prime	$\mathbb{N}_{F/K}$	$\mathbb{N}_{F/K}$
$\mathfrak{p}_2$	2	4	$\mathfrak{p}_{11}$	$\sqrt{-11}$	11
$\mathfrak{p}_3, \mathfrak{p}_3^*$	$(1+\sqrt{-11})/2$	3	$\mathfrak{p}_{23}, \mathfrak{p}_{23}^*$	$(9-\sqrt{-11})/2$	23
$\mathfrak{p}_3'$	π	3	$\mathfrak{p}_{31}, \mathfrak{p}_{31}^*$	$(5-3\sqrt{-11})/2$	31
$\mathfrak{p}_5, \mathfrak{p}_5^*$	$(3+\sqrt{-11})/2$	5	$\mathfrak{p}_{31}', \mathfrak{p}_{31}'^*$	$(5-3\sqrt{-11})/2$	31
$(3-\sqrt{-11})/2$	$((3-\sqrt{-11})/2)^2$	25			

From the ramification in F/K we get the relations $\mathfrak{p}_2^2=(2)$, $(\mathfrak{p}_3')^2=(\pi)$ and $\mathfrak{p}_{11}^2=(\sqrt{-11})$. We also find the following relations, checked by taking norms (and using congruences to distinguish between conjugate pairs) :

$$(\sqrt{-11}+\sqrt{\alpha})=\mathfrak{p}_3\mathfrak{p}_{11} \qquad (4+\sqrt{\alpha})=\mathfrak{p}_3^*\mathfrak{p}_5^*\mathfrak{p}_{31}$$

$$(1+\sqrt{\alpha})=(\mathfrak{p}_3^*)^2\mathfrak{p}_5 \qquad (2+\sqrt{\alpha})=\mathfrak{p}_3\mathfrak{p}_{31}'$$

$$(6+\sqrt{\alpha})=\mathfrak{p}_3'\mathfrak{p}_5^2\mathfrak{p}_{23}$$

Since K has class number 1, $\mathfrak{p}_a\mathfrak{p}_a^*$ is principal for any a. We conclude that the ideal class group of F has exponent at most 2, so Corollary 2 applies in this case also to show $Ш_3(\mathbb{Q})=0$ for $E^{(4)}$.

§3. The following Lemma is a simple exercise in Galois cohomology.

Lemma 4.

1) If $m\in\mathbb{Z}$ is odd, $S_m(\mathbb{Q})=S_m(K)^{\mathrm{Gal}(K/\mathbb{Q})}$.

2) If $\alpha,\beta\in\mathcal{O}$ are relatively prime then $S_{\alpha\beta}(K)=S_\alpha(K)\oplus S_\beta(K)$.

3) Writing $^-$ for conjugation by the nontrivial element of $\mathrm{Gal}(K/\mathbb{Q})$, $S_{\bar{\alpha}}(K)=\overline{S_\alpha(K)}$.

Proof of Theorem 1. Keep the notation of §1. Write π for the generator of $\mathfrak{p}$ with the property that the corresponding endomorphism of E reduces modulo $\mathfrak{p}$ to Frobenius. Recall that

$$S_\pi(K)=\ker\left[H^1(G_K,E_{\mathfrak{p}})\longrightarrow\bigoplus_v H^1(G_{K_v},E(\overline{K}_v))\right].$$

Define a larger group

$$S'_\pi(K) = \ker\left[H^1(G_K, E_\wp) \longrightarrow \bigoplus_{v\neq\wp} H^1(G_{K_v}, E(\overline{K_v}))\right],$$

so

$$\text{(3)} \qquad S_\pi(K) = \ker(S'_\pi(K) \longrightarrow H^1(G_{K_\wp}, E(\overline{K_\wp}))).$$

Write $\Delta = \mathrm{Gal}(F/K)$, let M be the maximal abelian p-extension of F unramified outside of the prime above $\wp$, and write $X = \mathrm{Gal}(M/F)$. By [4] and [9], $S'_\pi(K) \cong \mathrm{Hom}(X, E_\wp)^\Delta$.

Write Φ for the completion of F at the (unique) prime above $\wp$. Let U denote the units of Φ and $\mathcal{E}$ the closure of the global units of F in U. By class field theory we have an exact sequence

$$0 \longrightarrow \mathrm{Hom}(A, E_\wp) \longrightarrow \mathrm{Hom}(X, E_\wp) \longrightarrow \mathrm{Hom}(U/\mathcal{E}, E_\wp).$$

Takin Δ-invariants yields

$$\text{(4)} \qquad 0 \longrightarrow \mathrm{Hom}(A, E_\wp)^\Delta \longrightarrow S'_\pi(K) \longrightarrow \mathrm{Hom}(U/\mathcal{E}, E_\wp)^\Delta.$$

Next we need to determine $S_\pi(K)$ inside $S'_\pi(K)$. We have an exact sequence for Φ analogous to (1)

$$0 \longrightarrow E(\Phi)/\pi E(\Phi) \longrightarrow \mathrm{Hom}(G_\Phi, E_\pi) \longrightarrow H^1(G_\Phi, E(\overline{\Phi}))_\pi \longrightarrow 0.$$

We can identify Δ with $\mathrm{Gal}(\Phi/K_\wp)$ because $\wp$ is totally ramified in F/K. Now taking Δ-invariants gives (recall $|\Delta| = \mathbb{N}\wp - 1$)

$$\text{(5)} \qquad 0 \longrightarrow E(K_\wp)/\pi E(K_\wp) \longrightarrow \mathrm{Hom}(U, E_\wp)^\Delta \longrightarrow H^1(G_{K_\wp}, E(\overline{K_\wp}))_\wp \longrightarrow 0$$

since local class field theory shows $\mathrm{Hom}(G_\Phi, E_\wp)^\Delta \simeq \mathrm{Hom}(U, E_\wp)^\Delta$ and the inflation-restriction sequence shows

$$H^1(G_\Phi, E(\overline{\Phi}))^\Delta_\wp \simeq H^1(G_{K_\wp}, E(\overline{K_\wp}))_\wp.$$

Combining (3), (4) and (5) now yields an exact sequence for $S_\pi(K)$

$$\text{(6)} \qquad 0 \longrightarrow \mathrm{Hom}(A, E_\wp)^\Delta \longrightarrow S_\pi(K) \longrightarrow E(K_\wp)/\pi E(K_\wp).$$

Since $E(K_\wp)$ has no $\wp$-torsion ([5] Lemma 5), $\dim_{\mathcal{O}/\wp} E(K_\wp)/\pi E(K_\wp) = 1$. Thus $r \leq \dim_{\mathcal{O}/\wp} S_\pi(K) \leq r+1$ and the same holds for $\dim_{\mathbb{F}_p} S_p(\mathbb{Q})$ by Lemma 4.

To prove the second assertion of Theorem 1 we must now describe the image of $E(K_{\mathfrak{p}})/\pi E(K_{\mathfrak{p}})$ in $\mathrm{Hom}(U,E_{\mathfrak{p}})^{\Delta}$ in (5). When $\mathfrak{p}$ is a prime of ordinary reduction, and not anomalous, these are both one-dimensional $\mathcal{O}/\mathfrak{p}$-vector spaces and the problem is relatively simple. To deal with the other cases we will use elliptic units and the explicit reciprocity law of Wiles [13]. Let φ denote the map from $E(K_{\mathfrak{p}})$ to $\mathrm{Hom}(U,E_{\mathfrak{p}})^{\Delta}$ in (5). Fix a logarithm map λ (for our purposes this can be any isomorphism from $E(K_{\mathfrak{p}})/\pi E(K_{\mathfrak{p}})$ to $\mathfrak{p}/\mathfrak{p}^2$) and a generator u of $E_{\mathfrak{p}}$.

<u>Theorem 5</u>. There is a global unit $\epsilon\in\mathcal{E}$ such that for every $x\in E(K_{\mathfrak{p}})$,

$$\varphi(x)(\epsilon)= [(1-\pi^{-1})\Omega^{-1}L(E,1)\lambda(x)]u.$$

<u>Proof</u>. This is proved by combining the explicit reciprocity law of Wiles [13] with the fundamental computation of Coates and Wiles [5] relating elliptic units and $L(E,1)$. See for example [8], Theorems 3.5 and 6.8 (to derive the statement above for $p=3$ it is necessary to choose a larger group of elliptic units than was used in [8]; one needs to "extract 12th roots", which is possible by [6]).

Now suppose that $\Omega^{-1}L(E,1) \not\equiv 0 \pmod p$. Choose $s\in S_{\pi}(K)$ and let x be the image of s in $E(K_{\mathfrak{p}})/\pi E(K_{\mathfrak{p}})$ under (6). By (4), $\varphi(x)$ is 0 on all of $\mathcal{E}$, so in particular by Theorem 5 above,

$$(1-\pi^{-1})\Omega^{-1}L(E,1)\lambda(x) \equiv 0 \pmod{\mathfrak{p}}.$$

From our assumption on $L(E,1)$ we conclude that $\lambda(x)\equiv 0(\mathfrak{p}^2)$ so in fact $x\in\pi E(K_{\mathfrak{p}})$. Therefore by (6) $S_{\pi}(K)= \mathrm{Hom}(A,E_{\mathfrak{p}})^{\Delta} \simeq (\mathcal{O}/\mathfrak{p})^r$, so by Lemma 4 $S_p(\mathbb{Q}) \simeq \mathbb{F}_p^r$.

(*) p. 165. Partially supported by NSF grant DMS-8501937.

BIBLIOGRAPHY

[1] D. Bernardi, C. Goldstein, N. Stephens.- Notes p-adiques sur les courbes elliptiques. J. fur die reine und angew Math. 351 (1984), 129-170.

[2] B. Birch, H.P.F. Swinnerton-Dyer.- Notes on elliptic curves II. J. fur die reine und angew Math. 218 (1965), 79-108.

[3] J.W.S. Cassels.- Arithmetic on curves of genus 1. IV. J. fur die reine und angew Math. 211 (1962), 95-112.

[4] J. Coates.- Infinite descent on elliptic curves. In : Arithmetic and Geometry, papers dedicated to I.R. Shafarevich on the occasion of his 60th birthday, Vol. I. Progress in Math. 35, Boston : Birkhäuser (1983), 107-137.

[5] J. Coates, A. Wiles.- On the conjecture of Birch and Swinnerton-Dyer. Invent. Math. 39 (1977), 223-251.

[6] R. Gillard, G. Robert.- Groupes d'unités elliptiques. Bull. Soc. Math. France 107 (1979), 305-317.

[7] A. Odlyzko.- Lower bounds for discriminants of number fields II. Tohoku Math. J. 29 (1977), 209-216.

[8] K. Rubin.- Elliptic curves with complex multiplication and the conjecture of Birch and Swinnerton-Dyer. Invent. Math. 64 (1981), 455-470.

[9] K. Rubin.- Elliptic curves and $\mathbb{Z}_p$-extensions. Comp. Math. 56 (1985), 237-250.

[10] E. Selmer.- The diophantine equation $ax^3+by^3+cz^3=0$. Acta Math. 85 (1951), 203-262.

[11] G. Shimura.- On elliptic curves with complex multiplication as factors of the jacobians of modular function fields. Nagoya Math. J. 43 (1971), 199-208.

[12] N. Stephens.- The conjectures of Birch and Swinnerton-Dyer about $X^3+Y^3=DZ^3$. J. fur die reine und angew Math. 231 (1968), 121-162.

[13] A. Wiles.- Higher explicit reciprocity laws. Annals of Math. 107 (1978) 235-254.

[14] Numerical tables on elliptic curves, Table I. In : Modular Functions of One Variable IV, Lecture Notes in Math. 476, Berlin-Heidelberg-New York : Springer (1975), 76-113.

Karl RUBIN
Ohio State University
Columbus, OH 43210
U.S.A.

ON THE ARITHMETIC OF CONIC BUNDLE SURFACES*

Per SALBERGER

Introduction.

Let k be a perfect field and $\bar{k}$ an algebraic closure of k. Then a smooth projective k-surface X is said to be rational if $\bar{X} := \bar{k} \times_k X$ is integral and birational with $P^2_{\bar{k}}$. The aim of this paper is to study the arithmetic of some classes of rational surfaces with a conic bundle structure.

The first problem we are interested in concerns the Hasse principle. A class $\mathcal{C}$ of k-varieties over a number field k is said to satisfy the Hasse principle if for $X \in \mathcal{C}$: $X(k_v) \neq \emptyset$ at each completion k_v of $k \Rightarrow X(k) \neq \emptyset$. There are counterexamples to the Hasse principle for rational surfaces (cf. [CKS], [CSS, § 15]) but all the known counterexamples can be explained by an obstruction introduced by Manin (see [Ma_1], [Ma_2]). In particular, they satisfy $H^1(\bar{k}/k, \mathrm{Pic}\ \bar{X}) \neq 0$. In this paper we show that (cf. the theorem below) the Hasse principle holds for certain non-trivial classes of rational surfaces with $H^1(\bar{k}/k, \mathrm{Pic}\ \bar{X}) = 0$.

The second problem we are interested in concerns 0-cycles on X. Let $CH_0(X)$ be the group of 0-cycles modulo linear equivalence and $A_0(X)$ be the subgroup of $CH_0(X)$ of classes of 0-cycles of degree 0. The main tool in the study of $A_0(X)$ of (projective) rational surfaces is a characteristic homomorphism (cf. [Bl], [CS]) $\Phi_X : A_0(X) \longrightarrow H^1_{\acute{e}t}(k,T)$ where T is the k-torus defined by $T(\bar{k}) = \mathrm{Pic}(\bar{X}) \otimes_{\mathbf{Z}} \bar{k}^*$. It was proved in [Co] that Φ_X is injective if k is a number field and as a consequence of this that $A_0(X)$ is finite. Let Φ_{X_v} be the corresponding maps for $X_v := X \times k_v$ for the completions k_v at the places v of k. Then it was conjectured in [CS] that the natural map from coker Φ_X to $\coprod_{\text{all } v}$ coker Φ_{X_v} is injective for a number field k. If this conjecture is true one gets an effective method for determining the size of $A_0(X)$. Our main result is the following (cf. section 5).

<u>Theorem</u>. Let k be a number field and X/P^1_k a standard conic bundle (see (1.1)) with $K^2_X=4$. Suppose there is a field extension E of k of degree 4 with an intermediate field $k \subsetneq k' \subsetneq E$ such that X_E/P^1_E has a trivial generic fibre. Then :

(a) The Hasse principle and weak approximation hold for X if $H^1(\bar{k}/k, \mathrm{Pic}\ \bar{X})=0$ (that weak approximation holds for X means that $X(k)$, if non empty, is dense in $\prod_{v\in S} X(k_v)$ for any finite set S of places v of k).

(b) The conjecture above about $A_0(X)$ holds if $X(k)\neq\emptyset$.

The corresponding theorem for quadratic extensions E of k was proved in [CSS § 8] by Colliot-Thélène, Sansuc and Swinnerton-Dyer. They used universal torsors over X to reduce (a) and (b) to the arithmetic of certain intersections of two quadrics in P^7_k. Our approach is different since we use algebraic K-theory instead of universal torsors. But just as in [CSS] our problems are transformed into arithmetical questions about intersections of quadrics (see sections 4 and 5). The arithmetical problems one gets are harder but can be solved by means of section 13 of [CSS].

To reduce (b) to the arithmetic of such intersections of quadrics we could have used our main result in part [c] of our thesis [Sa_1], where we gave a concrete description of the homomorphism Φ_X for conic bundle surfaces over Severi-Brauer curves. This result cannot be used in the treatment of (a), however. Therefore, we work instead with restrictions of X/P^1_k to conic bundles U/A^1_k and with a homomorphism $\Phi_U : CH_0(U) \longrightarrow H^1(k,S)$ where S is the k-torus defined by $S(\bar{k})= \mathrm{Pic}(\bar{U}) \otimes_{\mathbf{Z}} \bar{k}^*$. The main goal of the first three sections is to prove (3.13) where we give a description of Φ_U in terms of norm groups associated to the quaternion algebra A corresponding to the generic fibre of U/A^1_k. This result replaces the main theorem of part [c] in [Sa_1]. To obtain it, we first construct an important diagram of K-groups associated to $\bar{U}/A^1_{\bar{k}}$ (see (1.8)). Then in section 3 we consider the Galois cohomology of the sequences in (1.8) and relate it to groups associated to A by means of results in section 2 (cf. (3.13)). In section 4, we use (3.13) to obtain a criterion (see (4.4)) for deciding when an element in $H^1(k,S)$ belongs to the image under the composition

$U(k) \longrightarrow CH_0(U) \longrightarrow H^1(k,S)$. Finally, in section 5 we apply this criterion in the case where k is a number field and prove the theorem stated above.

Let us mention that the results in this paper generalize the results in [Sa_3], where we assumed that E was biquadratic over k. Further, we would like to point out that very recently we have been able to generalize the theorem above to more general classes of conic bundle surfaces. These new results are also based on (3.13) in this paper.

We wish to thank J.-L. Colliot-Thélène for his careful reading of the paper.

1 - K-theory of conic bundles.

In this section we construct a diagram (1.8) which will be fundamental in our theory of Chow groups of conic bundle surfaces.

Let k be a perfect field of characteristic $\neq 2$ and $\bar{k}$ an algebraic closure of k. Further let $Y = A^1_k$, $\bar{Y} = Y \times_k \bar{k}$, K be the function field $k(Y)$ of Y, and $\bar{K}$ the function field of $\bar{Y}$.

(1.1) Definition : A conic bundle U/Y is a smooth connected k-surface with a proper k-morphism to Y, the generic fibre of which is a smooth conic over $k(Y)$. A standard conic bundle is a conic bundle U/Y such that the closed fibres U_y are isomorphic to conics in $P^2_{k(y)}$ for the residue field $k(y)$ at y.

The notion of standard conic bundles extends in an obvious manner to more general base curves and we will consider standard conic bundles $X/\tilde{Y} = P^1_k$ in section 5. But from now on U/Y will always denote a standard conic bundle over A^1_k. The following lemma is well known.

(1.2) Lemma : Let $U_{(y)} := U \times_Y \mathrm{Spec}\, O_{Y,y}$ for some closed point y of Y and let v_y be the discrete valuation : $K^* \longrightarrow\!\!\!\!\rightarrow \mathbf{Z}$ associated to $O_{Y,y}$. Then $U_{(y)}$ is isomorphic to $\mathrm{Proj}\,(O_{Y,y}[T_0,T_1,T_2]/(T_0^2-aT_1^2-bT_2^2))$ as a scheme over $\mathrm{Spec}\, O_{Y,y}$ for some $a,b \in K^*$ satisfying :

$$\text{(a)}\quad v_y(a) = v_y(b) = 0 \qquad \text{or} \qquad \text{(b)}\quad v_y(a)=0,\ \ v_y(b)=1.$$

(1.3) Remark : Note that in the case (b) we have a singular $k(y)$-point P_y of U_y defined by $T_0=T_1=0$. Further, if $\bar{a} = a + m_{Y,y}$ is a square in $k(y)^*$ then $U_y \backslash P_y$ is a disjoint union of two affine lines each of them defined over $k(y)$. If $\bar{a} \in k(y) \backslash k(y)^2$ then $U_y \backslash P_y$ is a disjoint union of two affine lines defined over $k(u(y)) := k(y)[T]/(T^2-\bar{a})$ and conjugated by the Galois group of $k(u(y))/k(y)$.

Now consider the Zariski sheaf $\mathcal{K}_2$ on $\bar{U} := U \times_k \bar{k}$ (cf. [Qu]). Then by obvious modifications of the arguments in [Bl, p. 44] we obtain :

(1.4) <u>Lemma</u> : (a) $H^0_{Zar}(\bar{U},\mathcal{K}_2) \xleftarrow{\sim} K_2(\bar{k})$

(b) $H^1_{Zar}(\bar{U},\mathcal{K}_2) \xleftarrow{\sim} \mathrm{Pic}(\bar{U}) \otimes_{\mathbf{Z}} \bar{k}^*$ (the isomorphism is induced by the symbol map $O^*_X \times O^*_X \longrightarrow \mathcal{K}_2$)

(c) $H^2_{Zar}(\bar{U},\mathcal{K}_2)=0$.

Let F (resp. $\bar{F}$) be the function fields of U (resp. $(\bar{U})$ and let U^r denote the set of points of codimension r on U. Then there is a complex :

(1.5) $$K_2(\bar{F}) \xrightarrow{\text{tame}} \coprod_{\bar{\gamma}\in\bar{U}^1} \bar{k}(\bar{\gamma})^* \xrightarrow{\text{div}} \coprod_{\bar{U}^2} \mathbf{Z}$$

where tame denotes the tame symbols (see [Mi]) and div the divisor maps. The maps in (1.5) can be identified with the boundary maps in Quillen's K-theory (cf. [Qu, p. 52] and [Ka, p. 645]). Hence by Gersten's conjecture (cf. [Qu, p. 48]) the complex (1.5) computes the Zariski cohomology of $\mathcal{K}_2$. Thus by (1.4) we get

(1.6) <u>Corollary</u> : (a) The divisor map in (1.5) is surjective.

(b) Let $\mathfrak{Z}$ be the kernel of the divisor map in (1.5). Then there is a short exact sequence of $\mathrm{Gal}(\bar{k}/k)$-modules :

$$0 \longrightarrow K_2(\bar{F})/K_2(\bar{k}) \xrightarrow{\text{tame}} \mathfrak{Z} \xrightarrow{\pi} \mathrm{Pic}(\bar{U}) \otimes_{\mathbf{Z}} \bar{k}^* \longrightarrow 0$$

where the restriction of π to $\coprod_{\bar{U}^1} \bar{k}^*$ is induced by the natural map from the group of Weil divisors $\mathrm{Div}(\bar{U})$ to $\mathrm{Pic}(\bar{U})$.

(1.7) <u>Notation</u> : By $k(u(y))$ we denote the $k(y)$-algebra of functions of the smooth locus of U_y which are constant on each connected component.

Thus $k(u(y)) = k(y)$ for closed points $y \in Y$ as in (1.2)(a) while $k(u(y)) = k(y)[T]/(T^2 - \overline{a})$ for $y \in Y^1$ as in (1.2)(b).

In the sequel we write $\overline{k}(y)$ for $\overline{k} \otimes_k k(y)$, $\overline{k}(u(y))$ for $\overline{k} \otimes_k k(u(y))$, and $\overline{K}(\gamma)$ for $\overline{K} \otimes_K K(\gamma)$. Further, we let $Pic_0(\overline{U}) := Ker\ (Pic(\overline{U}) \longrightarrow Pic\ (\overline{U}_K))$ for the restriction map to the generic fibre $\overline{U}_K$ of $\overline{U} \longrightarrow Y$.

(1.8) <u>Theorem</u> : There is a commutative diagram of $Gal(\overline{k}/k)$-modules with exact rows and columns :

$$
\begin{array}{ccccccccc}
 & & 0 & & 0 & & 0 & & \\
 & & \downarrow & \overset{K_2(\overline{K})/K_2(\overline{k})}{\swarrow\!\sim \qquad \searrow} & \downarrow & & \downarrow & & \\
0 & \longrightarrow & \coprod_{y \in Y^1} \overline{k}(y)^* & \longrightarrow & K_2(\overline{F})/K_2(\overline{k}) & \longrightarrow & K_2(\overline{F})/K_2(\overline{K}) & \longrightarrow & 0 \\
 & & \downarrow & & \downarrow{\scriptstyle tame} & & \downarrow{\scriptstyle tame} & & \\
0 & \longrightarrow & \coprod_{y \in Y^1} \overline{k}(u(y))^* & \longrightarrow & \mathfrak{Z} & \longrightarrow & \coprod_{\gamma \in U_K} \overline{K}(\gamma)^* & & \\
 & & \downarrow & & \downarrow & & \downarrow{\scriptstyle \sum_\gamma norm_\gamma} & & \\
0 & \longrightarrow & Pic_0(\overline{U}) \otimes_{\mathbb{Z}} \overline{k}^* & \longrightarrow & Pic(\overline{U}) \otimes_{\mathbb{Z}} \overline{k}^* & \longrightarrow & \overline{K}^* & & \\
 & & \downarrow & & \downarrow & \searrow \quad \overline{k}^* \hookrightarrow \quad \nearrow & \downarrow & & \\
 & & 0 & & 0 & & 0 & &
\end{array}
$$

where the maps from the second to the third columns are induced by restricting to the generic fibre (compare [CS (3.5)]) and where the injective maps in the first column and second row are the natural ones.

<u>Proof</u> : For the isomorphism used to define the map from $\coprod \overline{k}(y)^*$ to $K_2(\overline{F})/K_2(\overline{k})$, see [Mi, (2.3)]. The commutativity in the first two rows is clear since all non-trivial maps are given by tame symbols. To check the commutativity in the square starting with $\mathfrak{Z}$, one may restrict to $Div(\overline{U}) \otimes_{\mathbb{Z}} \overline{k}^*$ and calculate as in [CS p. 435] by means of (1.6)(b). The middle row is exact since $\overline{k}(u(y))^*$ is the

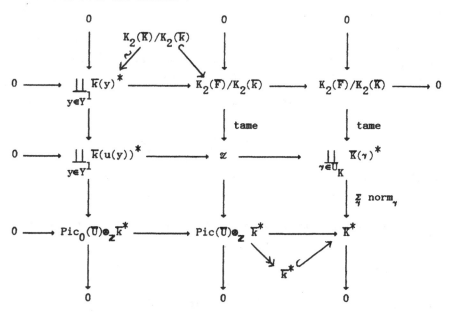

kernel of the divisor map from $\coprod_{\bar{\gamma}\in U_y^0} \bar{K}(\bar{\gamma})^*$ to $\coprod_{U_y^1} \mathbb{Z}$ (see (1.3)). For the exactness in the columns, see (1.6)(b) and [BT p. 31]. This completes the proof.

2 - The quaternion algebra associated to X_K.

We have seen in section 1 that X_K is isomorphic to a conic $T_0^2-aT_1^2-bT_2^2=0$, $a,b\in K^*$. Let $\{a,b\}_K$ be the quaternion algebra $A= K1+K\alpha+K\beta+K\alpha\beta$, $\alpha^2=a$, $\beta^2=b$, $\alpha\beta+\beta\alpha=0$. Note that A up to a K-algebra isomorphism depends only on X_K/K and not on the choice of a and b. We now turn to orders in A.

(2.1) Lemma. Let α, β be as above and $O_y := O_{Y,y}$ for some $y\in Y^1$. Then $\Lambda_y := O_y 1 + O_y\alpha + O_y\beta + O_y\alpha\beta$ is a hereditary order (see [Re]) if a, b, y satisfy (1.2)(a) or (1.2)(b). Moreover, if $S_y := \Lambda_y/\mathrm{rad}(\Lambda_y)$, then $S_y \xrightarrow{\sim} \{\bar{a},\bar{b}\}_{k(y)}$ in the case (a) and $S_y \xrightarrow{\sim} k(u(y))$ in the case (b).

Proof : One obtains immediately that $\mathrm{rad}(\Lambda_y)= m_y\Lambda_y$ in case (a), and $\mathrm{rad}(\Lambda_y)= m_y 1 + m_y\alpha + O_y\beta + O_y\alpha\beta$ in case (b). In both cases a simple calculation then gives that $\Lambda_y= \{\delta\in A : \delta\, \mathrm{rad}(\Lambda_y) \subseteq \mathrm{rad}(\Lambda_y)\}$. This implies that Λ_y is hereditary (see [Re (39.11-14)]).

For further connections between standard conic bundles and hereditary orders, see [Br].

(2.2) Definition : Let $y\in Y^1$ and S_y be as above. Then y is of type I if $S_y \simeq M_2(k(y))$, of type II if S_y is a skew field with $\dim_{k(y)}S_y=4$, of type III if S_y is a field with $\dim_{k(y)}S_y=2$ and of type IV if $S_y \simeq k(y) \oplus k(y)$ as a $k(y)$-algebra.

(2.3) Remark : It follows from (2.1) that X_y is a smooth conic over $k(y)$ with $X_y \simeq P^1_{k(y)}$ if y is of type I and with $X_y \simeq P^1_{k(y)}$ if y is of type II. If y is of type III or IV, then X_y is singular (cf. (1.3)).

Now recall that the Milnor ring $k_*(E)$ of a field E is the graded $\mathbb{Z}/2\mathbb{Z}$-algebra with generators $\ell(a)\in k_1(E)$, $a\in E^*$ and with relations $2\ell(a)=0$, $\ell(ab)= \ell(a)+\ell(b)$, $\ell(a)\ell(1-a)=0$ $a\neq 0,1$. It is well known that if E is of characteristic $\neq 2$ then $\ell(a)\ell(b) = \ell(a')\ell(b')$ if and only if $\{a,b\}_E \simeq \{a',b'\}_E$ as E-algebras.

Let $k_3(K) \xrightarrow{\partial_y} k_2(k(y)$ and $k_2(k(y))) \xrightarrow{\text{Galois}} Br(k(y))$ be the maps in [Mi]. Following Bloch [Bl], let ψ_y be the composition

$$K^* \xrightarrow{.\ell(a)\ell(b)} k_3(K) \xrightarrow{\partial_y} k_2(k(y)) \xrightarrow{\text{Galois}} Br(k(y))$$ where

$\{a,b\}_K$ corresponds to X_K and $.\ell(a)\ell(b)$ maps $c \in K^*$ to $\ell(a)\ell(b)\ell(c)$.

(2.4) <u>Lemma</u> : (a) Let y be of type I or IV and $c \in K^*$. Then $\psi_y(c)=0$.

(b) Let y be of type II and $c \in K^*$. Then $\psi_y(c)=0$ if $v_y(c)$ is even and $\psi_y(c) = [S_y]$ (the Brauer class of S_y) if $v_y(c)$ is odd.

(c) Let y be of type III and $c \in O^*_{Y,y}$. Then $\psi_y(c)$ is equal to the Brauer class of the cyclic algebra $(k(u(y))/k(y),\sigma,\bar{c})$, (cf. [Re]) where $\langle\sigma\rangle = Gal(k(u(y))/k(y))$ and $\bar{c}$ is the image of c in $k(y)^*$.

<u>Proof</u> : This follows from the definition of ∂_y [Mi p. 322] if we choose a,b satisfying (1.2)(a) resp. (1.2)(b); see the proof of lemma (3.15) in [Bl].

(2.5) <u>Lemma</u> : Let y be of type III and nr be the reduced norm from A to K (cf. [Re]). Then the natural map from $O^*_{Y,y}$ to $K^*/nr(A^*)$ is surjective. Further, there is a unique map sp_y from $K^*/nr(A^*) \longrightarrow k(y)^*/N(k(u(y))^*)$ such that

$$\begin{array}{ccc} K^*/nr(A^*) & \longrightarrow & k(y)^*/N(k(u(y))^*) \\ & \nwarrow \qquad \nearrow & \\ & O^*_{Y,y} & \end{array}$$ commutes

<u>Proof</u> : If we choose a,b as in (1.2)(b) then $b \in nr(A^*)$ by the definition of A. This proves the first statement since K^* is generated by b and $O^*_{Y,y}$. To prove the existence of sp_y it suffices to show that $\bar{c} \in N(k(u(y))^*)$ if $c \in O^*_{Y,y} \cap nr(A^*)$. But if $c \in nr(A^*)$, then $\psi_y(c)=0$ (see [Bl p. 51]). Now use (2.4)(c).

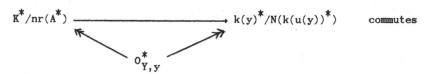

(2.6) Remark : It is easily seen from (2.4)(c) that

$$\begin{array}{ccc} K^* & \xrightarrow{\varphi_y} & Br(k(y)) \\ \downarrow & & \uparrow \\ K^*/nr(A^*) & \xrightarrow{sp_y} & k(y)^*/N(k(u(y))^*) \end{array}$$

commutes if y is of type III and the right hand map is the natural one (cf. [Re p. 263]).

We will also define sp_y for y of type IV. In that case (2.5) and (2.6) are trivially true since the norm map from $k(u(y))^*$ to $k(y)^*$ is surjective.

3 - The map $\Phi : CH_0(U) \longrightarrow H^1(\overline{k}/k, Pic(\overline{U}) \otimes_{\mathbf{Z}} \overline{k}^*)$.

From (1.6)(a) we obtain an exact sequence of $Gal(\overline{k}/k)$-modules :

(3.1) $$0 \longrightarrow \mathscr{Z} \longrightarrow \coprod_{\overline{\gamma}\in\overline{U}^1} \overline{k}(\overline{\gamma})^* \xrightarrow{div} \coprod_{\overline{U}^2} \mathbf{Z} \longrightarrow 0.$$

The cohomology sequence of (3.1) defines a homorphism $\coprod_{U^2} \mathbf{Z} \xrightarrow{\partial} H^1(\overline{k}/k,\mathscr{Z})$ which induces an isomorphism $i : CH_0(U) \xrightarrow{\sim} H^1(\overline{k}/k,\mathscr{Z})$ (see [Bl p. 45]). By composing this map with the map $H^1(\overline{k}/k,\mathscr{Z}) \longrightarrow H^1(\overline{k}/k,Pic(\overline{U}) \otimes_{\mathbf{Z}} \overline{k}^*)$ induced by (1.6)(b) we obtain a homomorphism $\Phi : CH_0(U) \longrightarrow H^1(\overline{k}/k, Pic(\overline{U}) \otimes_{\mathbf{Z}} \overline{k}^*)$ closely related to the homomorphism studied in [Bl] and [CS] for projective rational surfaces. The aim of this section is to give a very concrete interpretation of this map in terms of norm groups associated to A (cf. (3.13)).

(3.2) Definition : By $\mathscr{Z}_0$ we denote the inverse image of $Pic_0(\overline{U}) \otimes_{\mathbf{Z}} \overline{k}^*$ under the map $\pi : \mathscr{Z} \longrightarrow Pic(\overline{U}) \otimes_{\mathbf{Z}} \overline{k}^*$ (cf. (1.6)). By S (resp. S_0) we denote the k-tori defined by $S(\overline{k}) = Pic(\overline{U}) \otimes_{\mathbf{Z}} \overline{k}^*$ (resp. $S_0(\overline{k}) = Pic_0(\overline{U}) \otimes_{\mathbf{Z}} \overline{k}^*$).

From (1.8) we have the following three diagrams of $Gal(\overline{k}/k)$-modules (all the diagrams (3.3)-(3.8) are commutative with exact rows and columns).

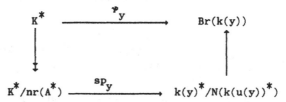

(3.3)
$$\begin{array}{ccccccccc}
0 & \longrightarrow & \mathcal{Z}_0 & \longrightarrow & \mathcal{Z} & \longrightarrow & \bar{k}^* & \longrightarrow & 0 \\
 & & \downarrow & & \downarrow{\scriptstyle \pi} & & \downarrow{\scriptstyle \mathrm{id}} & & \\
0 & \longrightarrow & S_0(\bar{k}) & \longrightarrow & S(\bar{k}) & \longrightarrow & \bar{k}^* & \longrightarrow & 0
\end{array}$$

(3.4)
$$\begin{array}{ccccccccc}
0 & \longrightarrow & \mathcal{Z}_0 & \longrightarrow & \mathcal{Z} & \longrightarrow & \bar{k}^* & \longrightarrow & 0 \\
 & & \downarrow & & \downarrow & & \downarrow & & \\
0 & \longrightarrow & K_2(\bar{F})/K_2(\bar{K}) & \longrightarrow & \coprod_{\gamma \in U_K} \bar{K}(\gamma)^* & \xrightarrow{\Sigma N_\gamma} & \bar{K}^* & \longrightarrow & 0
\end{array}$$

(3.5)
$$\begin{array}{ccccccccc}
 & & 0 & & 0 & & & & \\
 & & \downarrow & & \downarrow & & & & \\
0 & \longrightarrow & \coprod_y \bar{k}(y)^* & \longrightarrow & K_2(\bar{F})/K_2(\bar{k}) & \longrightarrow & K_2(\bar{F})/K_2(\bar{K}) & \longrightarrow & 0 \\
 & & \downarrow & & \downarrow & & \downarrow{\scriptstyle \mathrm{id}} & & \\
0 & \longrightarrow & \coprod_y \bar{k}(u(y))^* & \longrightarrow & \mathcal{Z}_0 & \longrightarrow & K_2(\bar{F})/K_2(\bar{K}) & \longrightarrow & 0 \\
 & & \downarrow & & \downarrow{\scriptstyle \pi|_{\mathcal{Z}_0}} & & & & \\
 & & S_0(\bar{k}) & \xrightarrow{\mathrm{id}} & S_0(\bar{k}) & & & & \\
 & & \downarrow & & \downarrow & & & & \\
 & & 0 & & 0 & & & &
\end{array}$$

By taking cohomology of (3.3)-(3.5) we deduce

(3.6)
$$\begin{array}{ccccccc}
k^* & \longrightarrow & H^1(\bar{k}/k,\mathcal{Z}_0) & \longrightarrow & H^1(\bar{k}/k,\mathcal{Z}) & \longrightarrow & 0 \\
\downarrow{\scriptstyle \mathrm{id}} & & \downarrow & & \downarrow & & \\
k^* & \longrightarrow & H^1(\bar{k}/k,S_0(\bar{k})) & \longrightarrow & H^1(\bar{k}/k,S(\bar{k})) & \longrightarrow & 0
\end{array}$$

(3.7)
$$\begin{array}{ccccccccc}
 & & k^* & \longrightarrow & H^1(\bar{k}/k,\mathcal{Z}_0) & \longrightarrow & H^1(\bar{k}/k,\mathcal{Z}) & \longrightarrow & 0 \\
 & & \downarrow & & \downarrow & & & & \\
\coprod_{\gamma \in U_K} K(\gamma)^* & \xrightarrow{\Sigma N_\gamma} & K^* & \longrightarrow & H^1(\bar{k}/k,K_2(\bar{F})/K_2(\bar{K})) & \longrightarrow & 0 & &
\end{array}$$

$$\begin{array}{ccccccc}
(3.8)\quad 0 \longrightarrow & H^1(\bar{K}/k,\mathscr{Z}_0) & \longrightarrow & H^1(\bar{K}/k,K_2(\bar{F})/K_2(\bar{K})) & \longrightarrow & \coprod_y \mathrm{Br}\, k(u(y)) \\
& \downarrow & & \downarrow & & \downarrow \mathrm{id} \\
0 \longrightarrow & H^1(\bar{K}/k,S_0(\bar{K})) & \longrightarrow & \coprod_y \mathrm{Br}\, k(y) & \longrightarrow & \coprod_y \mathrm{Br}\, k(u(y)).
\end{array}$$

To see that the first square in (3.8) commutes, note that we have an exact sequence of $\mathrm{Gal}(\bar{K}/k)$-modules :

$$(3.9)\qquad 0 \longrightarrow \coprod_y \bar{K}(y)^* \longrightarrow \mathscr{Z}_0 \longrightarrow K_2(\bar{F})/K_2(\bar{K}) \oplus S_0(\bar{K}) \longrightarrow 0$$

induced by (3.5) but where $\mathscr{Z}_0 \longrightarrow S_0(\bar{K})$ is given by $-\pi|_{\mathscr{Z}_0}$. Then consider the cohomology sequence of (3.9) and use the fact that $2\, H^1(\bar{K}/k,S_0(\bar{K}))=0$ (cf. (3.8)).

(3.10) <u>Remark</u> : It was shown in [CS p. 434] that $\mathrm{Im}(\Sigma\, N_y)= \mathrm{nr}(A^*)$ in (3.7). One thus gets an isomorphism

$$K^*/\mathrm{nr}(A^*) \xrightarrow{\sim} H^1(\bar{K}/k,\ K_2(\bar{F})/K_2(\bar{K})).$$

The following result was proved in [Bl (3.12)].

(3.11) <u>Lemma</u> : The map $\{\varphi_y\} : K^* \longrightarrow \coprod_y \mathrm{Br}\, k(y)$ in section 2 is equal to the composed map $K^* \longrightarrow H^1(\bar{K}/k,K_2(\bar{F})/K_2(\bar{K}) \longrightarrow \coprod_y \mathrm{Br}\, k(y)$ obtained from (3.7) and (3.8).

(3.12) <u>Notation</u> : $K^*_{dn} := \{\alpha\in K^*$ such that $v_y(\alpha)$ is even for each valuation $v_y : K^* \twoheadrightarrow \mathbb{Z}$ associated to a point $y\in Y$ of type II$\}$.

The notation K^*_{dn} (dn= divisorial norms) is motivated by the fact that an element α in K^* belongs to K^*_{dn} if and only if for each $y\in Y^1$ there is a unit α_y in $O_{Y,y}$, such that $\alpha/\alpha_y\in \mathrm{nr}(A^*)$.

We now come to the main theorem of this section.

(3.13) <u>Theorem</u> : There is a commutative diagram with exact rows :

$$\begin{array}{ccccccc}
k^* & \longrightarrow & K^*_{dn}/nr(A^*) & \longrightarrow & CH_0(U) & \longrightarrow & 0 \\
\downarrow{\scriptstyle id} & & \downarrow{\scriptstyle \{sp_y\}} & & \downarrow{\scriptstyle \Phi} & & \\
k^* & \longrightarrow & \prod k(y)^*/N(k(u(y))^*) & \longrightarrow & H^1(\bar{k}/k, S(\bar{k})) & \longrightarrow & 0
\end{array}$$

where the map $k^* \longrightarrow K^*_{dn}/nr(A^*)$ is the obvious one and where the product runs over all y of type III or IV.

<u>Proof</u> : From (2.4) and (3.11) it follows that K^*_{dn} is equal to the kernel of the composed map
$K^* \longrightarrow H^1(\bar{k}/k, K_2(F)/K_2(\bar{K})) \longrightarrow \coprod_y Br\ k(u(y))$ obtained from (3.7) and (3.8). Hence by (3.10) the first row of (3.8) induces an isomorphism $K^*_{dn}/nr(A^*) \xrightarrow{\sim} H^1(\bar{k}/k, \mathbb{Z}_0)$.

The second row in (3.8) defines an isomorphism $H^1(\bar{k}/k, S_0(\bar{k})) \xrightarrow{\sim} \prod k(y)^*/N(k(u(y))^*)$. Thus in view of (3.6) all that remains to do is to verify that the following diagram commutes :

$$\begin{array}{cccc}
 & K^*_{dn}/nr(A^*) & \xrightarrow{\{sp_y\}} & \prod k(y)^*/N(k(u(y))^*) \\
k^* \nearrow\!\!\!\searrow & \downarrow & & \downarrow \\
 & H^1(\bar{k}/k, \mathbb{Z}_0) & \longrightarrow & H^1(\bar{k}/k, S_0(\bar{k}))
\end{array}$$

The commutativity of the triangle follows from (3.7). To see that the square commutes, use (2.6), (3.11) and the commutativity of the first square in (3.8). This completes the proof.

In the sequel we will need some further properties of the diagram in (3.13). For each field extension k' of k we have a diagram associated to $U \times_k k'/Y \times_k k'$. We will use the fact that the diagrams in (3.13) are functorial with respect to such base extensions. This is easily seen from the functoriality of (3.3)-(3.5) (as diagrams of $Gal(\bar{k}/k)$-modules) under field extensions of k.

Now let t be an element of K^* such that $k[Y]= k[t]$ for the ring of regular functions $k[Y]$ on Y.

(3.14) <u>Proposition</u> : Let y be a k-point on Y defined by the equation $Ct+D=0$ for some C,D in k with $C\neq 0$. Then :

(a) $U_y(k)\neq\emptyset$ if and only if $Ct+D\in K^*_{dn}$.

(b) If P is a k-point on U_y then the class of $Ct+D$ in $K^*_{dn}/nr(A^*)$ is mapped to the class of P in $CH_0(U)$ in the diagram of (3.13) (this result should be regarded as a part of the theorem above).

<u>Proof</u> : (a) By the definition of K^*_{dn} we have $Ct+D\in K^*_{dn}$ if and only if y is of type I, III or IV. Hence (a) follows from (1.3) and (2.3).

(b) Let $Y_0= Y\setminus y$, $W= U\setminus U_y$, $\overline{Y}_0= \overline{K}\times_k Y_0$ and $\overline{W}= \overline{K}\times_k W$.

Then there is a complex :

$$(3.15) \qquad K_2(\overline{F}) \xrightarrow{\text{tame}} \coprod_{\overline{\gamma}\in\overline{W}^1} \overline{K}(\overline{\gamma})^* \xrightarrow{\text{div}} \coprod_{\overline{W}^2} \mathbf{Z}$$

which computes the Zariski-cohomology of $\mathcal{K}_2$ on $\overline{W}$ (cf. (1.5)). Hence if $\mathcal{Z}_{\overline{W}}$ is the kernel of the divisor map in (3.15) we obtain an isomorphism $CH_0(W) \xrightarrow{\sim} H^1(\overline{K}/k,\mathcal{Z}_{\overline{W}})$ just as for U by using the fact that $H^2_{Zar}(\overline{W},\mathcal{K}_2)= CH_0(\overline{W})=0$.

Next, let $H^1_{Zar}(\overline{U}_{\overline{K}},\mathcal{K}_2) \xrightarrow{\sim} \overline{K}^*$ be the isomorphism which we obtain from Gersten's conjecture (cf. [Qu p. 48]) and the last column of (1.8). Then by contracting possible exceptional curves in the fibres of $\overline{W}/\overline{Y}_0$ and by applying the results in [Bl p. 44] and [Sh] we obtain that the image of the composition $H^1_{Zar}(\overline{W},\mathcal{K}_2) \longrightarrow H^1_{Zar}(\overline{U}_{\overline{K}},\mathcal{K}_2) \xrightarrow{\sim} \overline{K}^*$ is equal to $\overline{K}[\overline{Y}_0]^*$.

We have, therefore, a commutative diagram of $Gal(\overline{K}/k)$-modules with exact rows (compare (3.4)) :

$$(3.16)\quad \begin{array}{ccccccccc}
0 & \longrightarrow & \mathcal{Z}_0 & \longrightarrow & \mathcal{Z} & \longrightarrow & \bar{K}^* & \longrightarrow & 0 \\
 & & \downarrow & & \downarrow & & \downarrow & & \\
0 & \longrightarrow & (\mathcal{Z}_{\bar{W}})_0 & \longrightarrow & \mathcal{Z}_{\bar{W}} & \longrightarrow & \bar{k}[Y_0]^* & \longrightarrow & 0 \\
 & & \downarrow & & \downarrow & & \downarrow & & \\
0 & \longrightarrow & K_2(\bar{F})/K_2(\bar{K}) & \xrightarrow{\text{tame}} & \coprod_{\gamma\in U_K} \bar{K}(\gamma)^* & \xrightarrow{\Sigma N_\gamma} & \bar{K}^* & \longrightarrow & 0
\end{array}$$

where $(\mathcal{Z}_{\bar{W}})_0$ is the kernel of the composition $\mathcal{Z}_{\bar{W}} \longrightarrow H^1(\bar{W},\mathcal{K}_2) \longrightarrow H^1(\bar{U}_{\bar{K}},\mathcal{K}_2)$. By taking cohomology and applying (3.8), (3.10) we get the following commutative diagram with exact rows :

$$(3.17)\quad \begin{array}{ccccccc}
k^* & \longrightarrow & H^1(\bar{K}/k,\mathcal{Z}_0) & \longrightarrow & CH_0(U) & \longrightarrow & 0 \\
\downarrow & & \downarrow & & \downarrow & & \\
k[Y_0]^* & \longrightarrow & H^1(\bar{K}/k,(\mathcal{Z}_{\bar{W}})_0) & \longrightarrow & CH_0(W) & \longrightarrow & 0 \\
\downarrow & & \downarrow & & & & \\
K^*/nr(A^*) & \longrightarrow & H^1(\bar{K}/k,K_2(\bar{F})/K_2(\bar{K})). & & & &
\end{array}$$

Now (b) follows from (3.17) since $k[Y_0]^*$ is generated by k^* and $Ct+D$ and since $Ker(CH_0(U) \longrightarrow CH_0(W))$ is generated by the linear equivalence class of P.

4 - <u>The image of the composition</u> $U(k) \longrightarrow CH_0(U) \xrightarrow{\Phi} H^1(k,S)$.

From now on we assume that our standard conic bundle $U \longrightarrow Y = A_k^1$ is the restriction of a conic bundle $X \longrightarrow \tilde{Y} = P_k^1$ to the complement of some smooth fibre $F := X_{y_\infty}$, $y_\infty \in \tilde{Y}(k)$. The aim of the section is to characterize the elements of $H^1(k,S)$ that are images under Φ of classes of $CH_0(U)$ containing effective cycles of degree one. We relate this question to the existence of rational points on certain intersections of quadrics.

We will write sp for the composition $K^* \longrightarrow K^*/nr(A^*) \xrightarrow{\{sp_y\}} \prod k(y)^*/N(k(u(y))^*)$ where y runs over the points of type III and IV. By $\Delta : \prod k(y)^*/N(k(u(y))^*) \longrightarrow Br\ k$

we will mean the homomorphism induced by the natural inclusions $k(y)^*/N(k(u(y))^*) \longrightarrow \mathrm{Br}\, k(y)$ and the corestriction map $\coprod \mathrm{Br}\, k(y) \xrightarrow{\Sigma \mathrm{cor}_y} \mathrm{Br}\, k$. Finally, we denote by $[a_F] \in \mathrm{Br}\, k$ the Brauer class of the quaternion algebra corresponding to $F = X_{y_\infty}$.

The following lemma is crucial for our method.

(4.1) Lemma : Let $Ct+D \in K^*$ be a linear polynomial with $c \neq 0$ and t as in (3.14). Then $Ct+D \in K^*_{dn}$ if and only if $\Delta(sp(Ct+D)) = [a_F]$.

Proof : Consider the map $\varphi_{y_\infty} : K^* \longrightarrow \mathrm{Br}\, k(y_\infty)$ defined as for $y \in Y$ (see sec. 2). Then since y_∞ is of type I or II we get $\varphi_{y_\infty}(Ct+D) = [a_F]$ by (2.4). Now recall that there is a complex :

$$k_3(K) \xrightarrow{\{\partial_y\}} \coprod_{y \in \tilde{Y}^1} k_2(k(y)) \xrightarrow{\Sigma N_y} k_2(k)$$

where the norm maps N_y are compatible with the corestriction maps $\mathrm{cor}_y : \mathrm{Br}\, k(y) \longrightarrow \mathrm{Br}\, k$. Thus $-\varphi_{y_\infty}(Ct+D) = \sum_{y \in Y} \mathrm{cor}_y(\varphi_y(Ct+D))$. Let y_0 be the k-point defined by $Ct+D=0$. Then by (2.4) and (2.6) we get $\sum_{y \in Y} \mathrm{cor}_y(\varphi_y(Ct+D)) = \Delta(sp(Ct+D)) + \varphi_{y_0}(Ct+D)$ with $\varphi_{y_0}(Ct+D) \neq 0$ if y_0 is of type II, and $\sum_{y \in Y} \mathrm{cor}_y(\varphi_y(Ct+D)) = \Delta(sp(Ct+D))$ otherwise. Summarizing, we have shown that $\Delta(sp(Ct+D)) = [a_F]$ if and only if y_0 is of type I, III or IV. This completes the proof (cf. (3.12)).

Now we investigate when an element $\eta \in \prod k(y)^*/N(k(u(y))^*)$, equals $sp(Ct+D)$ for some $Ct+D \in k[t]$ with $C \neq 0$. To do this, let $p(t)$ be the separable unitary polynomial whose irreducible factors correspond to the points $y \in Y$ of type III and IV, n its degree and R the semilocal ring $\cap O_{Y,y}$ defined by these points. Further, let Λ be a hereditary R-order in A which is ramified everywhere (i.e. such that $p(t)$ divides the discriminant $d(\Lambda/R)$, see [Re § 25]). Such orders clearly exist. We may take $\Lambda = \cap \Lambda_y$ for $O_{Y,y}$-orders Λ_y as in (2.1). It is also easy to see that Λ is unique up to conjugation by elements in A^* (compare [Sa$_2$ (3.1)]). Let $e_1, \ldots, e_m$ be elements of Λ whose classes in $\Lambda/\mathrm{rad}\,\Lambda$ form a basis of this k-vector space. From $\Lambda/\mathrm{rad}(\Lambda) = \sqcup k(u(y))$ for the points y of type III and IV we get $m=2n$. Given a polynomial $r(t) \in k[t]$ there exists a unique n-tuple of quadratic forms $\{Q_k\}$ $(k = 0, \ldots, n-1)$ with coefficients in k in variables U_i $(i = 1, \ldots, 2n)$ such that :

(4.2) $$r(t)\mathrm{Nrd}_{A/K}(\Sigma_i U_i e_i) \equiv \Sigma_k t^k Q_k \pmod{p(t)}.$$

Let $\tilde{V}$ be the subvariety of projective space P_k^{2n-1} with coordinates $(U_1,\dots,U_{2m})$ defined by $Q_k=0$ for $k=2,\dots,n-1$ and let $V\subset\tilde{V}$ be the open subset defined by $Q_1\neq 0$. Further, let $g : V \longrightarrow Y=\mathrm{Spec}\ k[t]$ be the k-morphism defined by $-Q_0/Q_1$.

(4.3) <u>Proposition</u> : Suppose $r(t)\in k[t]$ is relatively prime to $p(t)$ and suppose that $\Delta(\eta)=[a_F]$ for $\eta := sp(r(t))$. Then if $y_0\in Y(k)$ there exists $Ct+D\in m_{Y,y_0}$ with $\eta = sp(Ct+D)$ if and only if $y_0\in g(V(k))$.

<u>Proof</u> : We first observe that the map $\Lambda/\mathrm{rad}(\Lambda) \longrightarrow R/p(t)R$ induced by the reduced norm map $\Lambda \longrightarrow R$ coincides with the sum of the norm maps $k(u(y)) \longrightarrow k(y)$ formed by points of type III and IV. Hence by (4.2) a k-point y_0 belongs to $g(V(k))$ if and only if there exists $Ct+D\in m_{Y,y_0}$ such that $sp_y(r(t))=sp_y(Ct+D)$ for all points $y\in Y\backslash y_0$ of type III or IV. But if y_0 is of type III or IV then $Ct+D\in K^*_{dn}$ (cf. (3.12)) which in turn by (4.1) and the assumption that $\Delta(\eta)=[a_F]$ implies that $sp_{y_0}(r(t))=sp_{y_0}(Ct+D)$. This completes the proof.

Now suppose we are given an element $\epsilon\in H^1(k,S)$. Then we may choose $r(t)\in k[t]$ with $r(t)$ relatively prime to $p(t)$ so that $\eta=sp(r(t))$ is mapped to ϵ under the homomorphism in (3.13). By combining (3.13), (3.14), (4.1) and (4.3), we get :

(4.4) <u>Theorem</u> : Let $\epsilon, r(t), V, p(t)$ be as above for a standard conic bundle U/Y and let $y\in Y(k)$. Then there is a k-point P in U_y which is mapped to ϵ under the composition $U(k) \longrightarrow CH_0(U) \xrightarrow{\Phi} H^1(k,S)$ if and only if $\Delta(sp(r(t)))=[a_F]$ and $y\in g(V(k))$.

In order to apply the theorem when $n=4$, we will need :

(4.5) <u>Lemma</u> : Suppose $A=M_2(K)$, $p(t)=\Pi_{j=1}^4(t-c_j)$ $c_j\in k$ and $r(t)\in k[t]$ relatively prime to $p(t)$. Then with a suitable choice of coordinates $\tilde{V}$ is the subvariety of P_k^7 defined by the equations :

$$\Sigma_{j=1}^4 U_{2j-1}U_{2j}=0, \qquad \Sigma_{j=1}^4 c_j U_{2j-1}U_{2j}=0.$$

Proof : We may choose Λ to be $\begin{bmatrix} R & (p(t)) \\ R & R \end{bmatrix}$ and

$$e_{2j-1} = \begin{bmatrix} p_j(t) & 0 \\ 0 & 0 \end{bmatrix} \quad e_{2j} = \begin{bmatrix} 0 & 0 \\ 0 & p_j(t) \end{bmatrix} \quad \text{for } p_j(t) := p(t)/(t-c_j)$$

$j = 1,\dots,4$. Then
$r(t)\mathrm{nr}(\Sigma_{i=1}^{8} U_i' e_i) \equiv \Sigma_{j=1}^{4} r(c_j) U_{2j-1}' U_{2j}' p_j(t)^2 \equiv \Sigma_{j=1}^{4} r(c_j) p'(c_j) U_{2j-1}' U_{2j}' p_j(t)$ (mod $p(t)$). Hence by (4.2) we have :
$Q_3(U_1',\dots,U_8') = \Sigma_{j=1}^{4} r(c_j) p'(c_j) U_{2j-1}' U_{2j}'$
$Q_2(U_1',\dots,U_8') + (\Sigma_{j=1}^{4} c_j) Q_3(U_1',\dots,U_8') = \Sigma_{j=1}^{4} c_j r(c_j) p'(c_j) U_{2j-1}' U_{2j}'$.
Hence the lemma follows by taking $U_{2j-1} = r(c_j) U_{2j-1}'$, $U_{2j} = p'(c_j) U_{2j}'$.

(4.6) Proposition : Suppose $n=4$ and $r(t)$ relatively prime to $p(t)$. Then : (a) $\tilde{V}$ is a pure geometrically integral intersection of two quadrics in P_k^7 and $\tilde{V}$ is not a cone.

(b) V is smooth. The singular locus of $\tilde{V} \times_k \bar{k}$ consists of 8 points.

Proof : (a) This follows immediately from the previous lemma applied to $\tilde{V} \times_k \bar{k}$. See (1.1) and (1.11) in [CSS].

(b) A simple calculation shows that the singular locus of $\tilde{V} \times_k \bar{k}$ consists of the points where seven of the coordinates in (4.5) vanish. These points do not lie on $V \times_k \bar{k}$ since it is clear from the proof of (4.5) that there are $C_j \in \bar{k}$ such that $Q_1(U_1,\dots,U_8) = \Sigma_{j=1}^{4} C_j U_{2j-1} U_{2j}$ for the given coordinates.

The following result is also important in the applications of (4.4).

(4.7) Proposition : Suppose $n=4$, $r(t)$ relatively prime to $p(t)$ and that A is split by $K' := k'(t)$ for some quadratic extension k' of k. Then either $\tilde{V}$ contains a k-line or two disjoint skew projective spaces P_k^3, which are conjugate under the action of $\mathrm{Gal}(k'/k)$.

Proof : Suppose first that there is a point y of type IV of degree ≥ 2. Then we may choose $\overline{e_1}$ and $\overline{e_2}$ in $k(u(y))$ which are linearly independent over k and such that $N(U_1\overline{e}_1 + U_2\overline{e}_2) = 0$ under the norm map to $k(y)$ for any U_1, U_2 in k. Now note that $\overline{e_1}$ and $\overline{e_2}$ can be lifted to e_1 and e_2 in Λ such that $\mathrm{nr}(U_1 e_1 + U_2 e_2) \equiv 0$ (mod $p(t)$) for any linear combination over k. Hence by (4.2) the linear span of

e_1 and e_2 defines a line on $\tilde{V}$. It is also clear that we obtain a line on $\tilde{V}$ by a similar argument if we have two points of type IV. Therefore, by (4.8) below it only remains to treat the case where there are no points of type IV. But then Λ is unique and maximal (cf. (5.3), (39.14) and (40.5) in [Re]). Therefore $k' \subset \Lambda$ for an embedding $k'(t) \subset A$ (see [Re (5.3) and (12.8)]). Now let $e_j = t^{j-1}$, $e_{j+4} = \alpha t^{j-1}$, $j=1,\dots,4$ for an element α in Λ with $k' = k(\alpha)$ and $a := \alpha^2$ in k. Then $nr(\Sigma_{i=1}^{8} U_i e_i) = (\Sigma_{j=1}^{4} U_j t^{j-1})^2 - a(\Sigma_{j=1}^{4} U_{j+4} t^{j-1})^2$ which implies that $U_j = \alpha\, U_{j+4}$, $j=1,\dots,4$ and $U_j = -\alpha\, U_{j+4}$, $j=1,\dots,4$ define two disjoint skew projective spaces $P^3_{k'}$, which are conjugate under the action of $Gal(k'/k)$.

(4.8) Lemma : Let $X/\tilde{Y} = P^1_k$ be a standard conic bundle such that the quaternion algebra A associated to $X/\tilde{Y}$ is split by $k'(Y)$ for a quadratic extension k' of k. Then the sum $r = \Sigma[k(y):k]$ over points $y \in \tilde{Y}$ of type III is even.

Proof : y is of type III if and only if the maximal $O_{\tilde{Y},y}$-orders in A ramifies (see (2.1) and [Re (18.4)]). Hence by Mattson's non-commutative Hurwitz' formula (cf. [Ma], [VV]) we get that the genus of A is $r-3$ and this number is shown to be odd in [VV p. 42] for quaternion algebras as above.

5 - Conic bundle surfaces over number fields.

In this section we consider conic bundles $X/\tilde{Y} = P^1_k$ over number fields k. For some of these we verify a conjecture by Colliot-Thélène and Sansuc [CS] about the size of $A_0(X) := Ker(CH_0(X) \longrightarrow CH_0(\overline{X}))$. Further, we establish the Hasse principle and weak approximation for some conic bundle surfaces with $H^1(\overline{k}/k, Pic(\overline{X})) = 0$. An important rôle in the proofs is played by (4.4) which reduces the problems to the arithmetic of certain intersections of two quadrics in P^7_k. We can then apply the following deep result from [CSS § 13] :

(5.1) Theorem : Let k be a number field and let $\tilde{W} \subset P^7_k$ be a pure geometrically integral intersection of two quadrics which is not a cone. Assume $\tilde{W} \times k' \subset P^7_{k'}$, contains a line for a quadratic extension k' of k. Then the Hasse principle and weak approximation hold for any smooth open subvariety W of $\tilde{W}$.

Now consider completions k_v at places v of k. Then we denote by $\mathscr{C}_v$ the map $U(k_v) \longrightarrow H^1(k_v,S)$ corresponding to the composition $\mathscr{C}$: $U(k) \longrightarrow CH_0(U) \xrightarrow{\Phi} H^1(k,S)$. Further, if $\epsilon \in H^1(k,S)$ we denote by ϵ_v its image in $H^1(k_v,S)$.

(5.2) <u>Theorem</u> : Let $f : U \longrightarrow Y = A^1_k$ be a standard conic bundle as in the previous section over a number field k. Suppose that $\overline{U}/\overline{Y}$ has four degenerate fibres and that there are quadratic extensions k' of k and k'' of k' such that the generic fibre of $U \times k''/Y \times k''$ is isomorphic to $P^1_{k''(Y)}$. Then if $\epsilon \in H^1(k,S)$ and $\epsilon_v = \mathscr{C}(P_v)$ for suitable k_v-points on U at all places v of k we have :

(a) There is a k-point P on u with $\epsilon = \mathscr{C}(P)$.
(b) If we are given neigbourhoods N_v of P_v in the v-adic topologies for a finite set of places S then we may find a k-point P as above such that $P \in N_v$ for all v in S.

<u>Proof</u> : (a) Let $r(t)$, V be as in (4.4). Then by the injectivity in $\mathrm{Br}\, k \longrightarrow \coprod_{\text{all } v} \mathrm{Br}\, k_v$ and (4.4) applied to each $U \times k_v/Y \times k_v$ we get $\Delta(sp(r(t)) = [a_F]$. Hence by (4.4) and (5.1) it suffices to prove that $\tilde{V} \times k'$ contains a line. But $\tilde{V} \times_k k'$ is associated to the image ϵ' of ϵ in $H^1(k',S)$ in the same way as $\tilde{V}$ is associated to ϵ. Thus by (4.7) we see that $\tilde{W} = \tilde{V} \times_k k'$ contains a k'-line or two disjoint skew projective space π_1 and π_2 of dimension 3 over k'' and which are conjugate under $Gal(k''/k')$. But in the latter case we see from (5.1) that $\tilde{W}$ contains a k'-point P and then it is easy to check that the k'-line $\ell := \langle P,\pi_1\rangle \cap \langle P,\pi_2\rangle$ lies on the quadrics defining $\tilde{W}$.

(b) Since $\tilde{V} \times k'$ contains a line we obtain from (5.1) that weak approximation hold for V. Therefore, by (4.4) we may choose a k-point P with $\epsilon = \mathscr{C}(P)$ such that $f(P)$ is arbitrarily close to each $f_v(P_v)$, $v \in S$ for the k_v-morphisms $f_v : U \times k_v \longrightarrow Y \times k_v$ induced by f. We may then complete the proof as in [CSS III.9] by applying the implicit function theorem to suitable restrictions of f_v.

(5.3) <u>Theorem</u> : Let k be a number field and $X/\tilde{Y} = P^1_k$ a standard conic bundle (cf. (1.1)) with $K_X^2 = 4$ and $H^1(\overline{K}/k, \mathrm{Pic}\, \overline{X}) = 0$. Suppose there are quadratic field extensions k' of k and E of k' such that the generic fibre of $X_E/\tilde{Y}_E$ is trivial. Then the Hasse principle and weak approximation hold for X.

<u>Proof</u> : In order to apply (5.2), let $U = X\setminus F$ and $Y=\tilde{Y}\setminus y_\infty$ for a point $y_\infty \in \tilde{Y}(k)$ such that $F := X_{y_\infty}$ is smooth. Further, let $\mathrm{Pic}_c \overline{U} := \mathrm{Ker}(\mathrm{Pic}\,\overline{X} \to \mathrm{Pic}\,\overline{F})$. Then the intersection pairing $\mathrm{Pic}\,\overline{X} \times \mathrm{Pic}\,\overline{X} \longrightarrow \mathbf{Z}$ induces a perfect pairing $\mathrm{Pic}\,\overline{U} \times \mathrm{Pic}_c\overline{U} \longrightarrow \mathbf{Z}$. From this and class field theory (cf. [Ta]) we obtain a sequence :

$$H^1(k,S) \longrightarrow \coprod_{\text{all } v} H^1(k_v,S) \longrightarrow \mathrm{Hom}_{\mathbf{Z}}(Ш^1(k, \mathrm{Pic}_c\,\overline{U}), \mathbf{Q}/\mathbf{Z}).$$

Hence the theorem follows from (5.2) if we can prove that $Ш^1(k,\mathrm{Pic}_c\overline{U})=0$. To show this, we use the exact sequences :

$$(\mathrm{Pic}\,\overline{X})^G \longrightarrow (\mathrm{Pic}\,\overline{F})^G \longrightarrow H^1(G,\mathrm{Pic}_c\overline{U}) \longrightarrow H^1(G,\mathrm{Pic}\,\overline{X})$$

where $G = \mathrm{Gal}(\overline{k}/k)$ and $0 \longrightarrow \mathrm{Pic}\,X \longrightarrow (\mathrm{Pic}\,\overline{X})^G \longrightarrow \mathrm{Br}\,k \longrightarrow \mathrm{Br}\,X$ (cf. [Is_1 § 2]) for X and all $X \times k_v$. Then if $X(k_v)\neq\emptyset$ for all v we obtain canonical isomorphisms between $\mathrm{Ker}(H^1(G, \mathrm{Pic}_c\overline{U}) \longrightarrow H^1(G, \mathrm{Pic}\,\overline{X}))$, $\mathrm{Coker}(\mathrm{Pic}\,X \longrightarrow \mathrm{Pic}\,(\overline{X}_K)^G)$, and $\mathrm{Coker}\,(\mathrm{Pic}\,(X_K) \longrightarrow (\mathrm{Pic}\,(\overline{X}_K)^G)$. Therefore, $Ш^1(k,\mathrm{Pic}_c\overline{U})=0$ follows from the fact that $X_K \simeq P^1_K$ if the generic fibres of all $X \times k_v/\tilde{Y} \times k_v$ are trivial (cf. (2.2) and (3.5) in [Is_1]).

(5.4) <u>Remark</u> : Let X be as in (5.3) but not necessarily with $H^1(\overline{k}/k, \mathrm{Pic}\,\overline{X})=0$. Then it can be deduced from (5.2) that the Brauer-Manin obstructions (cf. [MT § 5]) are the only obstructions to the Hasse principle and weak approximation (see (2.2) and (3.2) in [Sa_3]). Complete proofs will appear elsewhere.

(5.5) <u>Theorem</u> : Let k be a number field and $X/\tilde{Y}=P^1_k$ a standard conic bundle with $K_X^2=4$ and $X(k)\neq\emptyset$. Suppose there are quadratic field extensions k' of k and E of k' such that the generic fibre of $X_E/\tilde{Y}_E$ is trivial. Then "conjecture A" (cf. [CS (4.5)] and the introduction) about $A_0(X)$ holds.

<u>Proof</u> : It is known (cf. [MT (3.5)]) that such surfaces X are k-unirational. Therefore, we may choose a point $y_\infty \in \tilde{Y}(k)$ for which

$F := X_{y_\infty} = P^1_k$. Then for $U = X \setminus F$ there is an exact sequence

$$(5.6) \qquad 0 \longrightarrow \mathbf{Z} \longrightarrow \operatorname{Pic} \bar{X} \longrightarrow \operatorname{Pic} \bar{U} \longrightarrow 0$$

where $1 \in \mathbf{Z}$ is mapped onto the class of $\bar{F}$ in $\operatorname{Pic} \bar{X}$. Now let T be the k-torus defined by $T(\bar{k}) = \operatorname{Pic}(\bar{X}) \otimes_{\mathbf{Z}} \bar{k}^*$. Then there is a commutative diagram :

$$(5.7) \qquad \begin{array}{ccccccc} & & A_0(X) & \xrightarrow{\sim} & CH_0(U) & & \\ & & \downarrow \Phi_X & & \downarrow \Phi & & \\ 0 & \longrightarrow & H^1(k,T) & \longrightarrow & H^1(k,S) & \longrightarrow & \operatorname{Br} k \end{array}$$

where the bottom row is induced by (5.6), Φ_X the homomorphism described in [Bl] and $A_0(X) \longrightarrow CH_0(U)$ the obvious map. By using (5.7) for X and $X_v = X \times_k k_v$ and the injectivity of $\operatorname{Br} k \longrightarrow \coprod_{\text{all } v} \operatorname{Br} k_v$ we obtain that conjecture A is equivalent to the following assertion :

(5.8) Let $\epsilon \in H^1(k,S)$. Then $\epsilon \in \operatorname{Im} \Phi$ if $\epsilon_v \in \operatorname{Im} \Phi_v$ for all places v of k.

This follows from (5.2) if we can prove that we have surjections $U(k_v) \longrightarrow CH_0(U \times k_v)$ for all places v of k. But $A_0(X \times k_v) \xrightarrow{\sim} CH_0(U \times k_v)$. Hence it suffices to show that each 0-cycle of degree zero on $X \times k_v$ is rationally equivalent to a cycle $P_v - \Omega_v$ where $P_v \in U(k_v)$ and $\Omega_v \in F(k_v)$. To see this, use [CC Th. C] and the fact that $\mathscr{C}_v(P_v) = 0$ for $P_v \in U(k_v)$ which are close to F in the v-adic topology (compare [CSS § 8 th. 6 (b)]). This completes the proof.

(5.9) <u>Corollary</u> : Let k be a number field and X a del Pezzo surface of degree 4 over k with $\operatorname{Pic} X \neq \mathbf{Z}$. Suppose that one of the 16 lines on $\bar{X}$ is defined over a field extension of degree 4 over k. Then :

(a) The Hasse principle and weak approximation hold for X if $H^1(\bar{k}/k, \operatorname{Pic} \bar{X}) = 0$.

(b) Conjecture A holds if $X(k) \neq \emptyset$.

<u>Proof</u> : We may assume that X is minimal since (a) and (b) are known for rational surfaces with $K_X^2 \geq 5$ (cf. [MT (7.1)]). But then there

are two standard conic bundles $X \longrightarrow \tilde{Y} = P^1_k$ and $X \longrightarrow \tilde{Z} = P^1_k$ if $X(k_v) \neq \emptyset$ at all v (cf. [Is_2 th. 1 and 5]). Further, the 16 lines on $\bar{X}$ coincide with the irreducible components of the degenerated fibres of $X_{\bar{K}}/\tilde{Y}_{\bar{K}}$ and $X_{\bar{K}}/\tilde{Z}_{\bar{K}}$. Now let ℓ be a line on $\bar{X}$ which is defined over a field E of degree 4 over k, and suppose that ℓ is mapped to a point P on $\tilde{Z}_{\bar{K}}$. Then since $X/\tilde{Z}$ is relatively minimal (cf. [Is_2 th. 5]) P is defined over a field k' with $k \subsetneq k' \subsetneq E$ (see (1.3)). It is also clear from the proof of [Is_2 th. 5] that ℓ provides as with a section of $X_E \longrightarrow \tilde{Y}_E$. Therefore, (a) and (b) follow from the theorems above.

(5.10) <u>Remark</u> : In [Ma_2 § 31] (see also [KST]) there is a classification of minimal del Pezzo surfaces of degree 4 into 19 types. From (5.9) we see that the Hasse principle and weak approximation hold for surfaces of type V and XIV. Note also that these surfaces are not k-birational to Châtelet-surfaces since they remain minimal after quadratic extensions of k (see [KST]).

(*) p. 175 Supported by NFR.

BIBLIOGRAPHY

[Bl] S. Bloch.- On the Chow groups of certain rational surfaces, Ann. Sc. Ec. Norm. Sup. (4) 14 (1981), 41-59.

[Br] J. Brzezinski.- Arithmetical quadratic surfaces of genus 0, I, Math. Scand. 46 (1980), 183-208.

[BT] H. Bass, J. Tate.- The Milnor ring of a global field, in Lecture Notes in Mathematics 342, Springer-Verlag, Berlin-Heidelberg-New York 1973.

[CC] J.-L. Colliot-Thélène, D. Coray.- L'équivalence rationnelle sur les points fermés des surfaces rationnelles fibrées en coniques, Compositio Math. 39 (1979), 301-332.

[CKS] J.-L. Colliot-Thélène, D. Kanevsky, J.-J. Sansuc.- Arithmétique des surfaces cubiques diagonales, to appear in Lecture Notes in Mathematics (ed. G. Wüstholz), Springer-Verlag.

[Co] J.-L. Colliot-Thélène.- Hilbert's theorem 90 for K_2, with application to the Chow groups of rational surfaces, Invent. Math. 71 (1983), 1-20.

[CS] J.-L. Colliot-Thélène, J.-J. Sansuc.- On the Chow groups of certain rational surfaces : a sequel to a paper of S. Bloch, Duke Math. J. 48 (1981), 421-447.

[CSS] J.-L. Colliot-Thélène, J.-J. Sansuc, Sir Peter Swinnerton-Dyer.- Intersections of two quadrics and Châtelet surfaces, J. reine angew. Math. 373 (1987), 37-107 and 374 (1987), 72-168 (cf. C.R. Acad. Sci. Paris : Série I, Math. 298 (1984), 377-380).

[Is_1] V.A. Iskovskih.- Rational surfaces with a pencil of rational curves, Mat. Sbornik 74 (1967), 608-638 (eng. transl : Math. USSR-Sbornik 3 (1967), 563-587).

[Is_2] V.A. Iskovskih.- Minimal models of rational surfaces over arbitrary fields, Izv. Ak. Nauk. SSSR Ser. Mat. 43 (1979), 19-43 (engl. transl : Math.-USSR Izv. 14 (1980), 17-39).

[Ka] K. Kato.- A generalization of local class field theory by using K-groups II, J. Fac. Sci. Univ. Tokyo, Sec. IA 27 Nc 3 (1980), 603-683.

[KST] B.E. Kunyavskii, A.N. Skorobogatov, M.A. Tsfasman.- Combinatorics and geometry of Del Pezzo surfaces of degree 4, Uspekhi Mat. Nauk 40 : 6 (1985), 145-46 (engl. transl. : in Russian Math. Surveys 40 : 6 (1985), 131-132).

[Ma_1] Yu.I. Manin.- Le groupe de Brauer-Grothendieck en géométrie diophantienne, in Actes du Congrès Intern. Math. (Nice, 1970), Gauthiers-Villars, Paris, 1971, tome 1, 401-411.

[Ma_2] Yu.I. Manin.- Cubic forms : algebra, geometry, arithmetic, second ed., North-Holland, Amsterdam-London, 1986.

[Mat] H. Mattson.- A generalization of the Riemann-Roch theorem Ill. J. Math. 5 (1961), 355-375.

[Mi] J. Milnor.- Algebraic K-theory and quadratic forms, Invent. Math. 9 (1970), 318-344.

[MT] Yu.I. Manin, M.A. Tsfasman.- Rational varieties : algebra, geometry, arithmetic, Uspekhi Mat. Nauk. 41 (1986) 43-94 (engl. transl in Russian Math. Surveys 41 : 2 (1986), 51-116).

[Qu] D. Quillen.- Higher algebraic K-theory I, in Lecture Notes in Mathematics 341, Springer-Verlag, Berlin-Heidelberg-New York 1973.

[Re] I. Reiner.- Maximal orders, Academic Press, London 1975.

[Sa_1] P. Salberger.- K-theory of orders and their Brauer-Severi schemes, Thesis, Dept. of Math., Univ. of Göteborg, 1985.

[Sa_2] P. Salberger.- Galois descent and class groups of orders, in Lecture Notes in Math. 1142, Springer-Verlag, Berlin-Heidelberg-New York 1985.

[Sa_3] P. Salberger.- Sur l'arithmétique de certaines surfaces de del Pezzo, C.R. Acad. Sci. Paris, Série I, Math. 303 (1986), 273-276.

[Sh] C. Sherman.- Some theorems on the K-theory of coherent sheaves, Comm. in Alg. 7 (1979), 1489-1508.

[Ta] J. Tate.- The cohomology groups of tori in finite Galois extension of number fields, Nagoya Math. J. 27 (1966), 709-719.

[VV] M. Van den Bergh, J. Van Geel.- Division algebras over function fields, Isr. J. Math. 52 (1985), 33-45.

Per SALBERGER
Dept. of Mathematics
Université de Paris-Sud
91405 ORSAY CEDEX

Modular Forms on Noncongruence Subgroups

A.J. SCHOLL

1 - Introduction.

In this paper I will describe some results and open problems connected with noncongruence subgroups of $PSL_2(\mathbb{Z})$. Most of these have their origins in the fundamental paper of Atkin and Swinnerton-Dyer [1]. Considering how much we know about congruence subgroups and the associated modular forms, it is remarkable how little we can say in the general case (to avoid cumbersome language I shall speak of "congruence modular forms" and "noncongruence modular forms"). The principal difficulty is the absence of a satisfactory theory of Hecke operators. For congruence subgroups, the Hecke operators not only provide a direct interpretation of Fourier coefficients of modular forms in terms of eigenvalues, but also furnish a link with arithmetic, essentially through the representation theory of adèle groups. Moreover, using the action of Hecke operators one can calculate congruence modular forms with relative ease. For a noncongruence subgroup it is possible to define the Hecke algebra using double cosets (as in Ch. 3 of [9]), but it seems difficult to exploit (I take this opportunity to correct the erroneous assertion to the contrary at the beginning of [6]); and there seem to be no good alternative computational devices with which to calculate noncongruence modular forms. This problem is discussed in detail in [1].

In recent years much progress has been made in the theory of congruence modular forms through arithmetic algebraic geometry. Fortunately many of the techniques are still applicable in the general case; the results described in §§ 3 and 4 make heavy use of them. However, until the last paragraph I shall avoid the language of abstract algebraic geometry wherever possible.

2 - Modular curves and fields of definition.

Let Γ be a subgroup of $PSL_2(\mathbb{Z})$ of finite index. It acts on the Poincaré upper half-plane H, and the quotient $U_\Gamma = \Gamma\backslash H$ is a noncompact Riemann surface; replacing H by $H^* = H \cup \mathbb{P}^1(\mathbb{Q})$ gives a compactification $X_\Gamma = \Gamma\backslash H^*$. Note that the modular function j defines a finite mapping

$$\pi : X_\Gamma \longrightarrow \mathbb{P}^1(\mathbb{C})$$

branched only in the three points $0, 1728, \infty$; it follows that X_Γ can be naturally given the structure of an algebraic curve over $\overline{\mathbb{Q}}$. A remarkable theorem of Belyi implies that the converse is true :

<u>Theorem</u> [2]. Every projective, nonsingular, irreducible curve over $\overline{\mathbb{Q}}$ is isomorphic to X_Γ for some Γ. More precisely, if V is a nonsingular irreducible curve over $\overline{\mathbb{Q}}$, then there is an open $U \subseteq V$ and a subgroup $\Gamma \leq PSL_2(\mathbb{Z})$ such that $U \simeq U_\Gamma$.

(At the suggestion of the referee we indicate how to deduce this from [2]. By the proof of Theorem 4 of [2] it is clear that for suitable U as above there exists a finite étale morphism $\phi : U \longrightarrow \mathbb{P}^1 - \{0,1,\infty\}$ defined over $\overline{\mathbb{Q}}$. Now since for $\Gamma = \Gamma(2)$ we have $U_\Gamma \simeq \mathbb{P}^1 - \{0,1,\infty\}$ the result follows). We just want to remark here that Belyi's ingenious proof gives a subset U such that the finite set $V-U$ is rather large; it would be interesting to have an intrinsic characterisation of all possible U_Γ.

Given a subgroup Γ, we would like to know something about X_Γ and its field of definition. In general a curve does not possess a best possible field of definition; the following discussion (implicit in [1], and described there in detail when X_Γ has genus zero) shows that we have enough additional structure to obtain a well-defined field, and a canonical model for X_Γ over it.

Consider, instead of X_Γ, the quadruple $\underline{X} = (X_\Gamma, \infty, \pi, q^{1/m})$. Here ∞ is the cusp of X_Γ at infinity, m denotes its width, π is the mapping above, and $q^{1/m}$ is the uniformiser $\exp(2\pi i z/m)$ at ∞. This quadruple is defined over $\overline{\mathbb{Q}}$ and has no nontrivial automorphisms. Therefore, if L_Γ denotes the fixed field of $\{\sigma \in Gal(\overline{\mathbb{Q}}/\mathbb{Q}) : \underline{X}^\sigma \simeq \underline{X}\}$, then $\underline{X}$ has a unique model over L_Γ.

In connection with modular forms there is a smaller field which it is useful to associate with Γ. If Γ is conjugated by $\begin{pmatrix} 1 & a \\ 0 & 1 \end{pmatrix}$, the quadruple $\underline{X}$ is replaced by

$$\underline{X}^{(a)} = (X_\Gamma, \infty, \pi, e^{2\pi i a/m} q^{1/m}).$$

Let K_Γ be the smallest field such that $\{\underline{X}^{(a)} : a=0,1,\ldots,m-1\}$ is mapped into itself by $Gal(\overline{\mathbb{Q}}/K_\Gamma)$. Then by Kummer theory we find that there exists some $\kappa \in L_\Gamma^*$ with $C = \kappa^m \in K_\Gamma$, such that $L_\Gamma = K_\Gamma(\kappa)$ and the quadruple $(X_\Gamma, \infty, \pi, \kappa q^{1/m})$ is defined over K_Γ. The field K_Γ depends only on the class of Γ under conjugation by elements of the form $\begin{pmatrix} 1 & * \\ 0 & 1 \end{pmatrix}$.

For a congruence subgroup L_Γ is abelian over $\mathbb{Q}$, and in fact for $\Gamma(N)$, $\Gamma_1(N)$ and $\Gamma_0(N)$ we have $K_\Gamma = L_\Gamma = \mathbb{Q}$. However it is clear from Belyi's theorem that there are no restrictions on the fields in general. Now denote by $M_k(\Gamma)$, $S_k(\Gamma)$ the spaces of holomorphic modular and cusp forms of weight k on Γ, respectively.

<u>Proposition</u> : Let k be an even positive integer. Then there are bases for $M_k(\Gamma)$ and $S_k(\Gamma)$ comprised of forms whose Fourier expansions have the shape

$$f(z) = \sum_{n \geq 0} a(n)\, e^{2\pi i n z/m} \tag{1}$$

where $b(n) = a(n)\kappa^{-n}$ belongs to K_Γ.

In fact rather more is true. We can associate to Γ a certain integer N, which is the product of a finite number of "bad" primes, including all primes dividing m and the norms of the numerator and denominator of C; the proposition then holds under the additional requirement that $b(n) \in O_{K_\Gamma}[1/N]$. It is certainly not true that we can impose the condition $a(n) \in O_{L_\Gamma}$, as the denominators of a noncongruence modular form can be unbounded - an example is $\Delta(z)^{1/5}\Delta(2z)^{4/5}$. What evidence there is seems, on the contrary, to suggest an affirmative answer to the following :

<u>Question</u>. If f is a modular form whose Fourier coefficients are algebraic integers, is f necessarily a congruence modular form ?

The problem of which primes can be "bad", and which can ramify in K_Γ, is discussed in [1].

3 - <u>ℓ-adic representations and congruences</u>.

Recall [3] that if f is a cusp form on $\Gamma_0(N)$ of weight k, and is a normalised eigenfunction of the Hecke operators with rational eigenvalues, then one can associate to f a strictly compatible system of ℓ-adic representations $\rho_\ell : \mathrm{Gal}(\overline{\mathbb{Q}}/\mathbb{Q}) \longrightarrow GL_2(\mathbb{Q}_\ell)$ with the following property : if p does not divide ℓN then ρ_ℓ is unramified at p, and

$$\det \rho_\ell(\mathrm{Frob}_p) = p^{k-1}$$
$$\mathrm{Tr}\, \rho_\ell(\mathrm{Frob}_p) = p^{th} \text{ Fourier coefficient of } f.$$

For noncongruence modular forms, analogous representations can be constructed which are related to Fourier coefficients, not by identities as above, but by congruences [6]. For simplicity, we assume $K_\Gamma=\mathbb{Q}$ (in general, one obtains representations of $\mathrm{Gal}(\overline{\mathbb{Q}}/K_\Gamma)$).

Let $d= \dim S_k(\Gamma)$. Then associated to $S_k(\Gamma)$ there is a system of representations $W_{k,\ell}$ of $\mathrm{Gal}(\overline{\mathbb{Q}}/\mathbb{Q})$ over $\mathbb{Q}_\ell$ of dimension 2d with the following properties :

a) $W_{k,\ell}$ is unramified away from ℓN (N as in § 2).

b) There is a nondegenerate alternating pairing

$$W_{k,\ell} \times W_{k,\ell} \longrightarrow \mathbb{Q}_\ell(-k+1).$$

c) Write $H_p(T)$ for the characteristic polynomial of Frob_p on $W_{k,\ell}$, and let

$$H_p(T)= \sum_{r=0}^{2d} A_r(p)T^{2d-r}, \quad A_0(p) = 1.$$

Then the coefficients $A_r(p)$ are integers independent of ℓ, and the roots of $H_p(T)$ have absolute value $p^{(k-1)/2}$.

d) For every p with $p>k-1$, $p|N$, every $f\in S_k(\Gamma)$ of the shape (1) above, every $\alpha\geq 1$, and every $n\geq 1$ with $p^\alpha|n$, we have

$$a(np^d)+A_1(p)a(np^{d-1})+\ldots+A_{2d}(p)a(n/p^d) \equiv 0 \bmod p^{(\alpha+1)(k-1)} \tag{2}$$

(slightly weaker congruences hold if $p \leq k-1$). The congruence means the following : recall $a(n)= b(n)\kappa^n= b(n)C^{n/m}$. By Hensel's lemma there is a unique m^{th} root γ of C^{p-1} in $\mathbb{Z}_p$ which is congruent to 1 mod p. If the left hand side of (2) is divided by $C^{n/m}$, the resulting expression can be written in terms of the $b(n)$'s and integral powers of $C^{(p-1)/m}$; so substituting γ for $C^{(p-1)/m}$ gives the congruence a meaning.

In the simplest case d=1 (as in the example in § 4 below) then (b) implies that $H_p(T)= T^2-A(p)T+p^{k-1}$, and (2) becomes

$$a(np) \equiv A(p)a(n) - p^{k-1}a(n/p) \bmod p^{(\alpha+1)(k-1)},$$

to be compared with the identity

$$a(np) = a(p)a(n) - p^{k-1}a(n/p)$$

for a normalised eigenform on a congruence subgroup.

In the congruence case, the properties of ρ_ℓ imply the Ramanujan conjecture $|a(p)| \le 2p^{(k-1)/2}$. But for a noncongruence subgroup, we obtain no archimedean relation between the roots of the polynomials $H_p(T)$ and Fourier coefficients. The best estimate known for the Fourier coefficients is $a(n) \ll n^{k/2-1/5}$, due to Rankin and Selberg [8].

One last point should be made; by considering subgroups of $SL_2(\mathbf{Z})$ instead of $PSL_2(\mathbf{Z})$ one obtains similar result for forms of <u>odd</u> weight $k \ge 3$ (apart from a complication in the field of definition - see the end of [6]). See also [10] for a somewhat different setting with $k=3$. It is however an open problem to find an analogue of $W_{k,\ell}$ if $k=1$ (as Deligne and Serre have done in the congruence case). It is not clear how such a representation, if it existed, would be characterised in terms of modular forms, since the congruences (2) are vacuous when $k=1$.

4 - <u>An example</u>. One of the simplest noncongruence subgroup is a certain subgroup Γ_{711} of $PSL_2(\mathbf{Z})$, investigated in [1]. It is the unique subgroup of index 9 with a cusp of width 7 at infinity, and inequivalent cusps of width 1 at ± 2. We have $m=7$, $K_\Gamma = \mathbf{Q}$, $\kappa = 2^{-1/7}$, and $N=14$. In the case $k=4$, we have $d = \dim S_4(\Gamma_{711}) = 1$, and so the congruences take the simple form mentioned above. The representations $W_{4,\ell}$ are 2-dimensional, unramified away from 14ℓ; and they have the same shape as one would expect of the representations attached to a cusp form of weight 4 on $\Gamma_o(2^a 7^b)$ for some a,b. In fact we have :

<u>Theorem</u>.

Let $g(z) = \frac{1}{8}\eta(z)\eta(2z)\eta(7z)\eta(14z)\ (14P(14z)-7P(7z)+2P(2z)-P(z))$

$$= \sum_{n\ge 1} B(n)e^{2\pi inz}$$

where $\eta(z)$ is Dedekind's η-function and $P(z) = 1-24\sum_{n\ge 1} e^{2\pi inz}$. Then $g(z)$ is a newform of weight 4 on $\Gamma_0(14)$ and $A(p) = B(p)$ for every prime $p \ne 2,7$.

It is easy to calculate the integers $A(p)$, at any rate for $p \ne 3$. Since the representations $W_{k,\ell}$ arise in algebraic geometry, one can apply the Lefschetz fixed point formula to give a simple closed formula for $A(p)$, which involves the numbers of points on various elliptic curves mod p (for a congruence subgroup this formula reduces to the Eichler-Selberg formula for the trace of the Hecke operator T_p). Thus I was able to find g and show that $A(p)=B(p)$ for all $p < 200$, $p \neq 2,3,7$. Now by Serre's effective version of Falting's method (see [5]) this is more than sufficient to prove the theorem. I am very grateful to Serre for explaining how to apply his method in this situation. Full details will appear elsewhere.

5 - Eisenstein series.

In § 3 we only considered cusp forms. However the congruence properties described there hold for noncusp forms as well, with appropriate modifications. At this point let us just say that the appropriate substitute for $W_{k,\ell}$ is a Galois representation $W^*_{k,\ell}$ which contains $W_{k,\ell}$; the quotient $W^*_{k,\ell}/W_{k,\ell}$ has dimension $\dim(M_k(\Gamma)/S_k(\Gamma))$ and the eigenvalues of an unramified Frobenius on it have absolute value p^{k-1}.

For a congruence subgroup of level N, the Hecke operators commute with the action of $\mathrm{Gal}(\bar{\mathbb{Q}}/\mathbb{Q}(\zeta_N))$ on $W^*_{k,\ell}$ and decompose it as a sum of $W_{k,\ell}$ and one-dimensional representations, each associated to an Eisenstein series. The congruences just mentioned are in this case completely subsumed in the wellknown explicit formulae for the Fourier coefficients of Eisenstein series.

For congruence subgroups $W_{k,\ell}$ need not be a direct factor of $W^*_{k,\ell}$ (we shall see an example shortly). We expect this to be closely related to algebraicity properties of Eisenstein series.

To be very specific, consider first the case $k=2$. Let ν be the number of inequivalent cusps of Γ, and for each cusp P_i $(1 \leq i \leq \nu)$ let Γ_i be its isotropy subgroup in Γ. Choose $\sigma_i \in PSL_2(\mathbb{R})$ such that $\sigma_i \Gamma_i \sigma_i^{-1} = \begin{pmatrix} 1 & * \\ 0 & 1 \end{pmatrix}$. Then the Hecke-Eisenstein series

$$G_i(z)= \lim_{s\to 0^+} \sum_{\gamma\in\Gamma_i\backslash\Gamma} \frac{1}{(cz+d)^2|cz+d|^s}, \qquad \sigma_i\gamma = \begin{pmatrix} * & * \\ c & d \end{pmatrix},$$

is a nonholomorphic modular form of weight 2 on Γ. The differences $G_i - G_j$ are holomorphic, and span a subspace of $M_2(\Gamma)$ complementary to $S_2(\Gamma)$.

Theorem [7]. Let n_i $(1\leq i\leq\nu)$ be integers such that $\sum n_i=0$. Then the Fourier coefficients of $\sum n_iG_i$ all belong to $\bar{\mathbb{Q}}$ if and only if the divisor $D=\sum n_iP_i$ has finite order in the Jacobian of X_Γ.

If the condition holds, so that $eD=(F)$ for some $e\geq 1$ and some function F, then $\Sigma\, n_i G_i$ is, up to a constant multiple, the logarithmic derivative of F. Using Belyi's theorem, it is easy to find Γ and $\{n_i\}$ such that D is of infinite order in the Jacobian of X_Γ.

Now the Galois modules $W_{2,\ell}$ and $W^*_{2,\ell}$ are respectively the first ℓ-adic cohomology groups of X_Γ and U_Γ, and we have an exact sequence

$$0 \longrightarrow H^1(X_\Gamma,\mathbb{Q}_\ell) \longrightarrow H^1(U_\Gamma,\mathbb{Q}_\ell) \longrightarrow \tilde{H}^0(\text{cusps},\mathbb{Q}_\ell(-1)) \longrightarrow 0$$

of $\mathrm{Gal}(\overline{\mathbb{Q}}/K_\Gamma)$-modules. The divisor of degree zero D defines a 1-dimensional subspace of $\tilde{H}^0$, invariant under an open subgroup G of $\mathrm{Gal}(\overline{\mathbb{Q}}/K_\Gamma)$, and by pullback an extension of $\mathbb{Q}_\ell(-1)$ by $H^1(X_\Gamma,\mathbb{Q}_\ell)$ (as G-modules). By the ℓ-adic analogue of the Abel-Jacobi theorem (see Ribet [5]) this extension splits if and only if D is of finite order. Combining this with the remark in the preceding paragraph, we see that $W_{2,\ell}$ need not be a direct factor of $W^*_{2,\ell}$.

In view of this, we would conjecture for arbitrary k :

Conjecture. $W_{k,\ell}$ is a direct factor of $W^*_{k,\ell}$ (as $\mathrm{Gal}(\overline{\mathbb{Q}}/K_\Gamma)$-modules) if and only if all Eisenstein series of weight k on Γ have algebraic Fourier coefficients.

One can formulate an analogue with $W^*_{k,\ell}$ replaced by a suitable mixed Hodge structure. The role of the Jacobian of X_Γ in the case $k=2$ should in the general case be played by a group arising in K-theory, but I only have a reasonable candidate for this when $k=3$; this will be considered elsewhere.

BIBLIOGRAPHY

[1] A.O.L. Artin, H.P.F. Swinnerton-Dyer.- Modular forms on noncongruence subgroups, Proc. Symp. Pure Math. AMS XIX (1971), 1-25.

[2] G.V. Belyi.- Galois extensions of a maximal cyclotomic field, Math. USSR Izv. 14 (1980), 247-256.

[3] P. Deligne.- Formes modulaires et représentations l-adiques, Sém. Bourbaki exp. 355, Lecture Notes in Math. 179 (1969), 139-172.

[4] R. Livné.- Paper to appear in proceedings of a conference on arithmetic algebraic geometry, Arcata 1985.

[5] K. Ribet.- Kummer theory and extensions of abelian varieties by tori, Duke Math. J. 46 (1979), 745-761.

[6] A.J. Scholl.- Modular forms and de Rham cohomology; Atkin-Swinnerton-Dyer congruences, Invent. Math. 79 (1985), 49-77.

[7] A.J. Scholl.- Fourier coefficients of Eisenstein series on noncongruence subgroups, Math. Proc. Camb. Phil. Soc. 99 (1986), 11-17.

[8] A. Selberg.- On the estimation of Fourier coefficients of automorphic forms, Proc. Symp. Pure Math. AMS VIII (1965), 1-15.

[9] G. Shimura.- Introduction to the arithmetic theory of automorphic functions, Iwanami Shoten and Princeton Univ. Press, 1971.

[10] J. Stienstra, F. Beukers.- On the Picard-Fuchs equation and the formal Brauer group of certain elliptic K3-surfaces, Math. Ann. 271 (1985), 269-304.

A.J. SCHOLL
Dept. of Mathematical Sciences
Science Laboratories
University of Durham
Durham DH1 3LE
ENGLAND

Liste des Conférenciers

7 octobre 1985	J. Oesterlé.- Courbes elliptiques modulaires.
14 octobre 1985	S. Friedberg.- Poincaré series and Kloosterman sums.
21 octobre 1985	J.-F. Michon.- Codes et courbes elliptiques.
4 novembre 1985	R. Greenberg.- Groupe de Mordell-Weil dans une tour de corps.
25 novembre 1985	P. Philippon.- Lemmes de zéros dans les groupes algébriques commutatifs.
2 décembre 1985	J. Igusa.- Complex powers and their functional equations.
9 décembre 1985	G. Frey.- Arithmetical applications of modular elliptic curves.
16 décembre 1985	C. Smyth.- The geometry of conjugate algebraic numbers.
6 janvier 1986	B. Venkov.- Le monstre et les formes modulaires.
13 janvier 1986	E. Fouvry.- Sur la conjecture d'Artin à propos des racines primitives.
20 janvier 1986	R. Gillard.- Croissance du p-nombre de classes dans certaines $\mathbb{Z}_\ell$-extensions.
27 janvier 1986	A. Scholl.- Modular forms and non-congruence subgroups.
3 février 1986	P. Debes.- Résultats récents sur le théorème d'irréductibilité de Hilbert.
24 février 1986	J.-P. Serre.- Représentations modulaires du groupe de Galois de $\overline{\mathbb{Q}}$ sur $\mathbb{Q}$.
3 mars 1986	F. Laubie.- Sur la ramification des extension infinies de corps locaux.
10 mars 1986	P. Salberger.- On the arithmetic of intersections of two quadrics.
17 mars 1986	A.M. Odlyzko.- New algorithms for computing the Riemann zeta function.
24 mars 1986	J.-M. Deshouillers.- Sur le problème de Waring.

14 avril 1986	L. Moret-Bailly.- Points entiers des variétés arithmétiques.
21 avril 1986	W.C. W. Li.- Gauss sums for quadratic extensions of a local field.
28 avril 1986	M. Laurent.- Nouvelle démonstration du théorème d'isogénie (d'après Chudnovski).
5 mai 1986	G. Rhin.- Approximants de Padé et mesures d'irrationalité effective.
12 mai 1986	A. Wiles.- Représentations ℓ-adiques.
	D. Masser.- Linear relations on algebraic groups.
26 mai 1986	W. Schmidt.- Nombre de solutions des équations de Thue.
	S. Lang.- La hauteur en théorie de Nevanlinna et en théorie des nombres.
2 juin 1986	C. Klingenberg.- Tate conjecture for Hilbert modular surfaces.
9 juin 1986	K. Rubin.- Courbes elliptiques et théorie d'Iwasawa.

ERRATUM A l'ARTICLE

Multiplication complexe dans les pinceaux de variétés abéliennes

par Yves ANDRE

paru dans le Séminaire de Théorie des Nombres de Paris 1984-85, Birkhäuser (pp. 1-22).

Page 1. Dernière ligne : lire "énoncé" au lieu de "exposé".

Page 3. Remplacer le second alinéa du paragraphe 2.3 par : "si G est sans pôle en 0, on sait qu'il existe une extension finie K' de K, une matrice $Y_G \in GL_n(K'((z)))$ et une matrice $C \in M_n(K')$, telles que $\partial Y_G = GY_G - Y_G C$ (avec $\partial = z\, d/dz$). Alors la matrice $X = Y_G z^C$ est une solution en 0 du système différentiel $(\partial - G)X=0$. De plus lorsque les différences des valeurs propres de $G(0)$ ne sont pas des entiers non nuls, on peut choisir $K'=K$, $C= G(0)$, et alors $Y_G \in GL_n(K[[z]])$, que l'on détermine complètement en imposant $Y_G(0)= id$".

6-7. Remplacer les termes en $n^2/p-1$ par n^4.

8. Ligne 16 : g n'est pas en général indépendant de θ, mais varie au plus linéairement en $h(\theta)$.

10. Paragraphe 4.3, ajouter : "$\varphi(s_0)=0$".

11. Ligne 3, ajouter : "après substitution $z \longrightarrow z^N$, N étant l'indice de ramification de φ au-dessus de 0"

Progress in Mathematics

Edited by:

J. Oesterlé
Departement des Mathematiques
Université de Paris VI
4, Place Jussieu
75230 Paris Cedex 05
France

A. Weinstein
Department of Mathematics
University of California
Berkeley, CA 94720
U.S.A.

Progress in Mathematics is a series of books intended for professional mathematicians and scientists, encompassing all areas of pure mathematics. This distinguished series, which began in 1979, includes authored monographs and edited collections of papers on important research developments as well as expositions of particular subject areas.

All books in the series are "camera-ready", that is they are photographically reproduced and printed directly from a final-edited manuscript that has been prepared by the author. Manuscripts should be no less than 100 and preferably no more than 500 pages.

Proposals should be sent directly to the editors or to: Birkhauser Boston, 675 Massachusetts Avenue, Suite 601, Cambridge, MA 02139, U.S.A.

1 GROSS. Quadratic Forms in Infinite-Dimensional Vector Spaces
2 PHAM. Singularités des Systèmes Différentiels de Gauss-Manin
3 OKONEK/SCHNEIDER/SPINDLER. Vector Bundles on Complex Projective Spaces
4 AUPETIT. Complex Approximation, Proceedings, Quebec, Canada, July 3–8, 1978
5 HELGASON. The Radon Transform
6 LION/VERGNE. The Weil Representation, Maslov Index and Theta Series
7 HIRSCHOWITZ. Vector Bundles and Differential Equations Proceedings, Nice, France, June 12–17, 1979
8 GUCKENHEIMER/MOSER/NEWHOUSE. Dynamical Systems, C.I.M.E. Lectures, Bressanone, Italy, June 1978
9 SPRINGER. Linear Algebraic Groups
10 KATOK. Ergodic Theory and Dynamical Systems I
11 BALSLEV. 18th Scandinavian Congress of Mathematicians, Aarhus, Denmark, 1980
12 BERTIN. Séminaire de Théorie des Nombres, Paris 1979–80
13 HELGASON. Topics in Harmonic Analysis on Homogeneous Spaces
14 HANO/MARIMOTO/MURAKAMI/OKAMOTO/OZEKI. Manifolds and Lie Groups: Papers in Honor of Yozo Matsushima
15 VOGAN. Representations of Real Reductive Lie Groups
16 GRIFFITHS/MORGAN. Rational Homotopy Theory and Differential Forms
17 VOVSI. Triangular Products of Group Representations and Their Applications
18 FRESNEL/VAN DER PUT. Géometrie Analytique Rigide et Applications
19 ODA. Periods of Hilbert Modular Surfaces
20 STEVENS. Arithmetic on Modular Curves
21 KATOK. Ergodic Theory and Dynamical Systems II
22 BERTIN. Séminaire de Théorie des Nombres, Paris 1980–81
23 WEIL. Adeles and Algebraic Groups

24 LE BARZ/HERVIER. Enumerative Geometry and Classical Algebraic Geometry
25 GRIFFITHS. Exterior Differential Systems and the Calculus of Variations
26 KOBLITZ. Number Theory Related to Fermat's Last Theorem
27 BROCKETT/MILLMAN/SUSSMAN. Differential Geometric Control Theory
28 MUMFORD. Tata Lectures on Theta I
29 FRIEDMAN/MORRISON. Birational Geometry of Degenerations
30 YANO/KON. CR Submanifolds of Kaehlerian and Sasakian Manifolds
31 BERTRAND/WALDSCHMIDT. Approximations Diophantiennes et Nombres Transcendants
32 BOOKS/GRAY/REINHART. Differential Geometry
33 ZUILY. Uniqueness and Non-Uniqueness in the Cauchy Problem
34 KASHIWARA. Systems of Microdifferential Equations
35 ARTIN/TATE. Arithmetic and Geometry: Papers Dedicated to I.R. Shafarevich on the Occasion of His Sixtieth Birthday, Vol. I
36 ARTIN/TATE. Arithmetic and Geometry: Papers Dedicated to I.R. Shafarevich on the Occasion of His Sixtieth Birthday, Vol. II
37 DE MONVEL. Mathématique et Physique
38 BERTIN. Séminaire de Théorie des Nombres, Paris 1981–82
39 UENO. Classification of Algebraic and Analytic Manifolds
40 TROMBI. Representation Theory of Reductive Groups
41 STANELY. Combinatories and Commutative Algebra
42 JOUANOLOU. Théorèmes de Bertini et Applications
43 MUMFORD. Tata Lectures on Theta II
44 KAC. Infinite Dimensional Lie Algebras
45 BISMUT. Large Deviations and the Malliavin Calculus
46 SATAKE/MORITA. Automorphic Forms of Several Variables Taniguchi Symposium, Katata, 1983
47 TATE. Les Conjectures de Stark sur les Fonctions L d'Artin en s = 0
48 FRÖHLICH. Classgroups and Hermitian Modules
49 SCHLICHTKRULL. Hyperfunctions and Harmonic Analysis on Symmetric Spaces
50 BOREL, ET AL. Intersection Cohomology
51 BERTIN/GOLDSTEIN. Séminaire de Théoire des Nombres. Paris 1982–83
52 GASQUI/GOLDSCHMIDT. Déformations Infinitesimales des Structures Conformes Plates
53 LAURENT. Théorie de la Deuxième Microlocalisation dans le Domaine Complèxe
54 VERDIER/LE POTIER. Module des Fibres Stables sur les Courbes Algébriques Notes de l'Ecole Normale Supérieure, Printemps, 1983
55 EICHLER/ZAGIER. The Theory of Jacobi Forms
56 SHIFFMAN/SOMMESE. Vanishing Theorems on Complex Manifolds
57 RIESEL. Prime Numbers and Computer Methods for Factorization
58 HELFFER/NOURRIGAT. Hypoellipticité Maximale pour des Opérateurs Polynomes de Champs de Vecteurs
59 GOLDSTEIN. Séminarie de Théorie des Nombres, Paris 1983–84
60 PROCESI. Geometry Today: Giornate Di Geometria, Roma. 1984
61 BALLMANN/GROMOV/SCHROEDER. Manifolds of Nonpositive Curvature
62 GUILLOU/MARIN. A la Recherche de la Topologie Perdue
63 GOLDSTEIN. Séminaire de Théorie des Nombres, Paris 1984–85
64 MYUNG. Malcev-Admissible Algebras
65 GRUBB. Functional Calculus of Pseudo-Differential Boundary Problems

66 Cassou-Noguès/Taylor. Elliptic Functions and Rings and Integers
67 Howe. Discrete Groups in Geometry and Analysis: Papers in Honor of G.D., Mostow on His Sixtieth Birthday
68 Robert. Antour de L'Approximation Semi-Classique
69 Faraut/Harzallah. Deux Cours d'Analyse
70 Adolphson/Conrey/Gosh/Yager. Number Theory and Diophantine Problems: Proceedings of a Conference at Oklahoma State University